HANDS ON
ENERGY, INFRASTRUCTURE
AND RECYCLING

PRACTICAL INNOVATIONS FOR A SUSTAINABLE WORLD

HANDS ON
ENERGY, INFRASTRUCTURE AND RECYCLING

Compiled by

Emma Judge

Published by ITDG Publishing
103–105 Southampton Row, London WC1B 4HL, UK
www.itdgpublishing.org.uk

© ITDG Publishing 2002

First published in 2002

Introduction and editing by Kim Daniel

ISBN 1 85339 488 2

All rights reserved. No part of this publication may be reprinted or reproduced or utilized in any form or by any electronic, mechanical, or other means, now known or hereafter invented, including photocopying and recording, or in any information storage or retrieval system, without the written permission of the publishers.

A catalogue record for this book is available from the British Library.

This document is an output from a project funded by the UK Department for International Development (DFID) for the benefit of developing countries. The views expressed are not necessarily those of DFID.

ITDG Publishing is the publishing arm of the Intermediate Technology Development Group. Our mission is to build the skills and capacity of people in developing countries through the dissemination of information in all forms, enabling them to improve the quality of their lives and that of future generations.

Typeset by J&L Composition Ltd, Filey, North Yorkshire
Printed in Great Britain

Contents

Preface	vii
Acknowledgements	xi
Introduction	xv

1 Power without destruction — 1
  Burning biogas — 2
  Capturing coal-bed methane — 7
  Micro-hydro power — 10
  Water current turbines — 15
  Harnessing wave energy — 20
  Wind power — 24
  Windpumps — 31
  Solar thermal power — 35
  Wood pulp for power — 39

2 Energy-efficient living — 43
  Reducing household fuel consumption — 44
  Energy-efficient lighting — 48
  Solar lanterns — 52
  Solar home systems — 56
  Schools energy-saving project — 60
  The Solar House — 63
  Urban energy policy — 67
  Passive House standards — 72

3 Recycling a valuable resource — 77
  Household sorting of domestic waste — 78
  Informal recycling of waste materials — 82
  Community waste collection — 85
  Automated recycling of drinks containers — 88
  Cash from cans — 92
  Accessories from inner tubes — 95
  From rags to handmade paper — 99

Paper from algae and other wastes	102
Rehabilitating water hyacinth	106
Fuel from plastic waste	110
Shipping containers as building materials	114
4 Transport for the future	**119**
Intermediate means of transport	120
Bicycle hire	124
Electric three-wheelers	127
Self-service car rental	132
Smart cars	135
Hydrogen power	138
Hybrid power	141
Bioengineering to prevent landslides	147
5 Building a safe environment	**155**
Improved traditional housing	156
Low-cost concrete housing	161
Earthen architecture	164
Cost-effective school buildings	169
Hurricane-resistant roofing	173
Earthquake-proof housing	178
Cyclone-resistant health centres	181
Small-scale brickmaking	186
Making lime cements	191
Reducing traffic noise	195
Bibliography	**199**
Appendix	**215**

Preface

'This programme inspired me to start a rural development programme in a very poor and underdeveloped area here' Dr R.K. Biswas, Calcutta, India

Hands On – It Works is a multi-media project which combines TV, radio, the web and the printed word in a unique initiative that emerged from letters received by the Director of Television Trust for the Environment (TVE)[1], Robert Lamb, in response to programmes on Earth Report which promoted practical 'hands on' initiatives. These letters echoed the demand expressed by correspondence received on an annual basis by ITDG's Technical Enquiry Service for more empowering information on appropriate technologies, and a partnership was born between ITDG and TVE. The initiative has led to the largest postbag received in response to programmes in TVE's 15-year history and has catapulted both agencies into the new multi-media age with a product which is ahead of its time.

In a typical week, the Hands On 'strand within a strand' which goes out as part of TVE's series, Earth Report, gets more reaction from viewers than all the BBC World output over the same seven-day period. By any criteria, this is an extraordinary achievement for TVE and the agencies that have supported Hands On and Earth Report.

By the time this publication is on the shelves, Hands On will have been produced and broadcast to 167 countries on the satellite, terrestrial and cable stations carrying the news and documentary channel. A recent breakthrough has been the sale of the Hands On – Earth Report series for transmission on the National Geographic Channel. This will ensure that the second package of Hands On is available to 85 million homes in 111 countries in 16 languages. Moreover, the take-up means that Hands On – Earth Report is now associated with two of the four most prestigious and most watched global factual programming services.

Daily feeds via the China Radio International (CRI) satellite service to over 60 TV stations in China have added a further 300 million homes to the outreach. Though it already goes out in Japan via Direct

TV, NHK is also likely to take up some of the 20 programmes. Courtesy of a three-year project due to be funded by the GEF and the European Commission, to begin in 2001, are broadcasts in Spanish and Portuguese to Latin America. In the offing are sales to Zee TV (India), UPC (pan Europe), Turkish TV and others. This builds on TVE's previous success in sales to national broadcasters in the French-speaking countries (via Odyssey), Spain, Germany, Ireland, the UK and Australia.

Not only have the responses to the programmes increased over the life of the project to a figure now approaching 35 000 hits per programme (excluding the letters, e-mails and faxes) there is now firm evidence that delivery of the *Hands On* multi-media package of programmes and back-up information are catalysing new initiatives; that governments and business are changing their policies; that NGOs are using clips or features for awareness raising; and that other agencies, like the United Nations Food and Agriculture Organization (FAO), are approaching *Hands On* to cover stories with a topical spin.

> 'I have never been one to respond in writing to a TV programme before, but this time I really wanted to say something ... this programme, Gone Fishing, got me thinking because I could see the tangible results' Kerry Hart, New Zealand. [Gone Fishing includes several examples of fisheries initiatives, which are described in Chapter 3 of *Hands On: Food, Water and Finance.*]

As take-up for the second package of twelve *Hands On* programmes (comprising 60 five-minute features) looks set to outstrip the success of the first series in increasing awareness of environmentally sound development among a global general public, *Hands On* remains the *only series* on world television dedicated exclusively to dissemination of empowering information which may inspire technology transfer. Measurements of success are not based on audience ratings alone but the true gauge is the hundreds of letters, faxes and e-mails received, the spread of countries reached and the increase of hits on the web site to 35 000, and rising, per programme.

In addition to these indicators are unsolicited reports from several of the entrepreneurs involved in the programmes about the uptake and dissemination of the technologies featured. Examples include: Trevor Field, the entrepreneur who developed the 'Play pump' in South Africa [included in *Hands On – Food, Water and Finance*], who rang TVE to reveal that thanks to *Hands On* he had received a World Bank Grant to disseminate the technology further; TOMRA, the company behind 'Cashing In' [published in *Hands On – Energy, Infrastructure and*

Recycling as 'Automated recycling of drinks containers'], a technology that encourages recycling using a Reverse Vending Machine, made a CD-ROM out of the feature and disseminated it to all of its offices worldwide; the Philippine windpump manufacturer thanked us for helping promote his business to people throughout the region; and the success of the 'Safa Tempo' ['Electric three-wheelers', again in *Hands On – Energy, Infrastructure and Recycling*] feature remains one of *Hands On's* greatest achievements as it led to a change of government policy in Nepal, and a ban on diesel polluting three-wheelers and ultimately a reduction in greenhouse gases.

Another measure recognising the programmes' potential for inspiration may be drawn from the support *Hands On* has received from organizations such as FAO and the International Fund for Agricultural Development (IFAD). Each of these was particularly anxious to maximize the multi-media nature of the project and encourage responses from viewers – in the case of FAO, this has led to a direct link with a small grants programme which provides potential entrepreneurs with the chance to raise money for start-up projects.

> *'This programme helps a lot of people in poor and undeveloped countries'* Dr So Myint Aung, Myanmar.

While the target audiences for the programmes remain entrepreneurs, local NGOs, women's groups and farmers, many of these will be reached only if policy makers are also made aware of the potential of environmentally sound development and the equally important benefits in terms of poverty reduction. *Hands On* has therefore tried to maximize its impact with special screenings at conferences: in The Hague at an international water conference; in Brighton at a Sustainable Energy Fair; and in Washington at the World Bank Rural Development Week. Use of *Hands On* by development professionals in training sessions and in public fora has also helped to further promote the series and its message that skills and know-how can be adapted and transferred to further sustainable development. The publication of these two books of compilations of the back-up material will also enable dissemination of the information beyond the life of the project. And there is no question but that the printed back-up information has been key to the success of the initiative, helping entrepreneurs (see below) realize the full potential of some of the technologies featured.

Keeping a check on the impact of a worldwide programme like *Hands On* is often difficult and relies on information submitted by partner organizations such as this remarkable story from China –

'*After viewing the Biogas programme, I was very impressed by the ideas presented. I made a presentation to Dr Si Zhizhong of the Canadian Embassy's Project Development Office . . . our project was approved and we formally signed a contract to construct 124 eight cubic metre biogas tanks with US$16 300 in aid . . . we estimate that each household can increase their income by US$100 a year and cut wood by 5400 kg, save 1.5 tonnes of coal pa, increase grains by 10%, increase vegetable yields by 20% and on average add three more pigs to their stock.*'

The aforementioned examples are clear evidence that the 'Hands On' project has helped to increase the awareness of environmentally sound development and inspire technology transfer and confirm the findings of the interim evaluation which concluded that:

'Hands On *is already increasing awareness among a number of target audiences: policy and decision makers in the North and South who can influence change and encourage sustainable development practice; entrepreneurs and business people interested in trying to implement ideas; and individuals who want to find out more or use the material to influence others (e.g. teachers). The preponderance of responses from manufacturers and small-scale entrepreneurs should be noted as the clearest indicator of likely take-up*'. Dario Pulgar

A final thanks should go to the Department for International Development (DFID) who provided core support to the *Hands On* project at a time when multi-media was a new concept; and to the United Nations Environment Programme and WWF, who have been constant in their backing for *Earth Report* without which *Hands On* would never have been possible.

Janet Boston
Television Trust for the Environment

---

* The Television Trust for the Environment (TVE) is an international non-profit organization working to raise awareness of environment, development, health and human rights issues through the media. It is committed to the distribution of films and multi-media packages to broadcasters, NGOs and extension agencies in the North and the South. It has gained international recognition for its pioneering work.

Acknowledgements

Intermediate Technology Development Group (ITDG) would like to thank Bob Spencer for his endless support and vast knowledge that contributed to the technical information, and Janet Boston for her innovation and enthusiasm in producing the *Hands On* television series.

ITDG would like to acknowledge the following. For Chapter 1: the Biogas Support Programme for providing the original material on biogas technology (Burning biogas); UNDP for providing the original material on the development of coal-bed methane resources in China (Capturing coal-bed methane); ITDG Peru, in particular Saul Ramirez, and Thropton Energy Services for providing the original information on river turbines for battery charging (Water current turbines); Wavegen and Ocean Power Delivery for providing the original material on wave power (Harnessing wave energy); Alexis Belonio, from the Central Philippine University, for his help in providing the original material on wind energy (Windpumps); Pilkington Solar International for the original material and pictures of solar thermal technology (Solar thermal power); the Austrian Energy Agency and the Austrian Federal Ministry of Science and Transport for providing the original information on biomass as a source of renewable energy (Wood pulp for power).

For Chapter 2: Hugh Allen for providing the original manual and pictures of the Kenya Ceramic Jiko (Reducing household fuel consumption); the Beijing Energy Efficiency Centre for providing the original information on the China Green Lights Programme (Energy-efficient lighting); ITC for permission to reproduce information and photographs of the solar lantern (Solar lanterns); Shell Renewables and Conlog (Pty) Limited for providing the original material on the Solar Power House™ system (Solar home systems); the members of the Fifty-Fifty Project, particularly Wolfgang Thiel, for their help in providing the original notes on the project and Hilary Dunne, from the Lawrence Sheriff School in Rugby, for translating the original material (Schools energy-saving project); Oxford Brookes University and Dr Susan Roaf who provided the original material on the Oxford Solar

House (The Solar House); the Environment Protection Department in the City of Freiburg and Solar-Fabrik for providing the original material on solar energy (Urban energy policy); the Cepheus project and the Austrian Energy Institute for providing the original material on the cost-efficient passive houses (Passive House standards).

For Chapter 3: the Municipality of Fredericia for providing the information on the refuse system in Fredericia (Household sorting of domestic waste); the Water, Engineering & Development Centre (WEDC) at Loughborough University, in particular Mansoor Ali, Mariëlle Snell and Andy Cotton, for providing the original information on solid waste management (Informal recycling of waste materials); TOMRA, and in particular Caroline Quinn, for providing the original material on reverse vending machines (Automated recycling of drinks containers); Prolata and Alcan Aluminium Limited for providing the original material and the pictures of the recycling of aluminium cans (Cash from cans); Cartiera Favini, in particular Marnie Campagnaro and Clemente Nicolucci, for providing the original material on environmentally friendly papermaking in Italy (Paper from algae and other wastes); the Prisons Water Hyacinth Team for producing the original material on the Water Hyacinth Art-Craft Training Project (Rehabilitating water hyacinth); the China Aerospace Great Wall Group, the Lan Ye Corporation and the Beijing Energy Efficiency Centre for providing the original materials on the conversion of waste plastics to liquid fuels (Fuel from plastic waste); BP Community Affairs for the information they provided on the conversion of shipping containers (Shipping containers as building materials).

For Chapter 4: Y-Tech Innovation Centre for providing the original material on the White Bicycle System in Amsterdam (Bicycle hire); Liselec, in particular Jacques Mollard and Jean-Michel Couturier, for providing all the original information on electric car rental (Self-service car rental); the Smart Car Information Centre for providing the original material on the Smart Car (Smart Cars); Ludwig-Bölkow-Systemtechnik GmbH for providing the original material on the hydrogen bus and fuel cells (Hydrogen power); Volvo, in particular Michael Borg and Lars-Ake Weimar, for providing the original material on the Environmental Concept Bus and Truck (Hybrid power); the Transport Research Laboratory, Old Wokingham Road, Crowthorne, Berkshire RG45 6AU, United Kingdom, and His Majesty's Government of Nepal, who prepared the original material on bioengineering in Nepal (Bioengineering to prevent landslides).

For Chapter 5: Dr Mohsen Aboutorabi, from the Birmingham School of Architecture, who wrote the original paper on Low Cost Housing

Projects for Townships of South Africa (Low-cost concrete housing); CRATerre-EAG, in particular Hugo Houben, for providing the original material and photographs of earthen architecture (Earthen architecture); the Department for International Development (DFID), the British Council, the Government of Andhra Pradesh and the Government of India for their support of the project on cost-effective technologies for primary school construction in Andhra Pradesh, and Roger Bonner and P.K. Das for providing the original material and pictures on the Andhra Pradesh Primary Education Project (Cost-effective school buildings); the Construction Resource and Development Centre in Jamaica for providing the original material on the Retrofit Programme and the National Housing Trust (Hurricane-resistant roofing); Paul Brown for providing the original article on using bamboo as a building material in Colombia (Earthquake-proof housing); Sudipto Mukerjee, P.K. Das, Ms. Peu Banerjee Das and Roger Bonner for providing the original material on effective low-cost building technologies in India (Cyclone-resistant health centres); ITDG's Building Materials Panel, particularly Michael Wingate, for their help in designing the Chegutu lime kiln, and Lotte Reimer of Ove Arup & Partners, PO Box 984, Harare, Zimbabwe who prepared the original notes on the kiln with the assistance of Kelvin Mason and Peter Tawodzera (Making lime cements); Wilma Bouw Engineering, in particular Roel Slagter, for providing the original material on the noise walls (Reducing traffic noise).

Introduction

Sustainability is now accepted as a key factor in any development project, as the problems of environmental pollution and depletion of resources affect both North and South, threatening the livelihoods, quality of life and health of the majority of people. Practical solutions to these problems, as well as to the poverty that persists in far too many parts of the globe, are being implemented by individuals and whole communities, by small businesses and large organizations, and by government and non-governmental agencies – often with financial rewards for environmentally friendly practices.

Examples of such solutions to local or global problems have been taken from two remarkable series of television programmes, which feature the work of agencies, entrepreneurs and communities all around the world in the fields of sustainable enterprise and appropriate technology. The examples that are included in this book and in the companion volume, *Hands On – Food, Water and Finance*, were selected because they describe a wide range of affordable and replicable technologies to combat common problems, and because they take into account the impact on the environment, as well as on the people, of the changes being introduced.

This volume includes examples of initiatives in the areas of power generation, energy-efficient living, recycling, transport, and building, while its companion volume covers water and sanitation, farming and food production, finance, and health.

The purpose of the two publications is to facilitate technology transfer from South to South or from North to South. The books provide a collection of ideas and approaches that work, and signpost sources of further information, so that development professionals can identify solutions that are appropriate to the area they are in, and find out more. Many of the innovations are likely to be of interest in the North, but the primary aim is to provide a source book of inspiration and practical ideas for those working in the South.

The challenges described in this book include satisfying the need for safe, comfortable living accommodation, ideally with access to power for lighting, refrigeration and a radio – usually a problem only for the poorest members of society; and organizing the disposal of rubbish without contributing to pollution in the local environment – a requirement for most people in any community to some extent.

In some spheres, there are very real differences between rural and urban needs. In the rural areas of developing countries, people are much less likely to have access to electricity from a national grid and so, far from reducing their power consumption, would like the facility to increase it in pursuit of a livelihood. Urban dwellers are more likely to be affected by traffic noise, emissions and congestion, by a build-up of rubbish or a shortage of space. Rural people have a greater need for transport to gain access to essential goods and services – though, as elsewhere, vehicles that do not add to global pollution would be the preferred solution.

Building techniques that optimize the use of local resources and reduce the need for fuel – whether for heating or air conditioning, as described in Chapter 2, or during the manufacture of materials and building construction, as described in Chapter 5 – are applicable to North and South, rural and urban requirements. Where there are more specific needs, as in areas that are prone to natural disasters – and such areas are becoming more common as a result of climate change – building design becomes critical. Traditional building materials and forms have been developed in India to increase the cyclone resistance of a number of new health centres, and in Colombia to enable housing to withstand earthquake damage.

Such innovations can cost very little, where waste materials are used to save resources – for example, chopping up rags to make paper in India, or using boiler ash to fire bricks in Zimbabwe. In other cases, the reduction in power or materials used can lead to considerable savings, which serves as an incentive for people to adopt more efficient practices. There is an example from Germany of a number of schools actively reducing their energy and water use, and receiving half the money saved in this way. The monetary gain almost certainly convinced the individual schools to participate, but it was found that the students are more interested in the corresponding reduction in carbon monoxide emissions, and so this aspect is emphasized for their benefit.

More commonly the initial costs are high, because of the research or capital expenditure required. Often it takes several years for any impact to be seen, as for example when using bioengineering to stabilize roads or when growing bamboo for use in building. Government agencies and development organizations, with resources at their disposal in terms of both purchasing power and information, can do much to help a community to implement sustainable development initiatives, but it is important to ensure that targets allow for this development phase, and also that the intended beneficiaries are involved from the beginning, if interventions are to succeed. Active participation by local people makes it more likely that the activity will be sustained and that any regulations will be monitored and enforced by the people themselves.

In Kenya, the Maasai people have been forced by land subdivision to abandon their nomadic tradition and so need more durable and permanent housing, but improvements are possible only with the approval and cooperation of the women whose work it is to build the houses. Subtle but effective changes to roof height and materials, and the amount of light and ventilation, have improved the living environment and enabled the collection of rainwater while reducing the maintenance demands – thereby increasing the women's free time without challenging their status as homemakers. In South Africa, low-cost accommodation to rehouse the residents of informal settlements was developed successfully only after the owners of the new houses were involved in decision making and trained to participate in the building and maintenance.

Life-changing innovations can be extremely simple. The adaptation of a rickshaw for refuse collection in Bangladesh is a good example – the streets are too narrow for the corporation trucks, so a local businessman has organized production of the rickshaws and pays sweepers to clean the streets. Householders are charged a small fee for the removal of their rubbish. The streets are now cleaner and safer, partly because residents are less likely to leave rubbish on the streets.

The use of a waste product such as manure for power or heat generation is another simple yet effective step. Wastes from the power industry can serve a useful purpose as a fuel or component of building materials. In other cases the technology is complex, as in most of the approaches to power generation, and those involved in the changes would therefore benefit from training and advice from those who have experience in the field. Whether it is a simple 'Why didn't I think of that?' or a complex, long-term undertaking, this book provides a wealth of material to assist the process of problem solving.

Within each chapter the examples of practical solutions are grouped thematically – although, inevitably, projects touch on many different areas, particularly in a book like this where the chapters' subject matter overlaps, and so they could have been presented in other combinations with different emphases. As far as possible they have been arranged from the simplest, lowest-input solutions to the largest, most complex. The chapters contain contact details for the individuals or organizations who were involved in the work described, or who advised ITDG about the technology, as well as other organizations in related fields. The bibliography contains references and details of other publications, by chapter, that will help readers to build a fuller picture of the technologies and options available.

Chapter One
Power without destruction

In many parts of the developing world there is no electricity grid, and diesel fuel, where available, can be expensive or of poor quality. There are numerous renewable sources of energy, however, one or more of which can usually be accessed in remote areas in order to charge batteries and so provide power for lighting, refrigeration and communications equipment, or to raise water or generate heat. Renewable energies have the added advantage of being free of climate-changing emissions. Initial costs tend to be high, but if finance is available the lower running costs or the ability to work after nightfall help to repay the investment.

Biogas provides a useful alternative to diminishing supplies of wood for cooking in Nepal, and has also been used to generate heat and electricity for farm buildings in Germany. One of the constituents of biogas, methane, is also present in coal deposits, where it must be captured or ventilated to the atmosphere in order to prevent explosions. As methane has much the same effect on the atmosphere as carbon dioxide, it makes sense to capture it for power generation, as in the example from China.

Areas with wet climates can benefit from water power without the need to install a large dam. In Peru, the power from micro-hydro schemes in the Andean highlands area has enabled health, education and other services to be developed, while in the jungle areas water current turbines have been installed. In coastal Scotland, wave power helps to meet growing energy needs while also contributing to sea defences and providing habitats for commercially valuable fish.

Windy climates offer the possibility of power generation on a local scale. Wind turbines are mainly installed in developing countries for battery charging or for water pumping. In The Philippines, where there is monsoon all year round, windpumps supply domestic water, irrigate farms and vegetable gardens, provide drinking water for livestock and supply water to petrol stations.

Solar energy is in plentiful supply in hot countries, and may also be a viable option in colder climates. Here we have examples of solar thermal technologies that are suitable for remote power supply and smaller energy needs.

Finally, there are still areas like those described in Austria, where managed woodland is used to produce timber and wood pulp, leaving large amounts of waste bark and other residues that can be used for electricity production and heating.

Burning biogas

In Nepal, there is a seemingly endless demand for fuelwood, which is used mainly for cooking. The pressure on forests has reached record levels, and deforestation, soil erosion and landslides occur frequently. Recognizing the urgent need to find an alternative source of fuel, the Nepalese Government, the Netherlands Development Organization and local banks set up the Biogas Support Programme to promote household biogas plants using animal dung as an available and appropriate fuel. As a result, biogas is being produced in family units for domestic purposes, such as cooking and lighting.

When it is used in appropriate burners, biogas gives a clean, smokeless blue flame that is ideal for cooking. For lighting, biogas is used in specially designed lamps which give a light similar to kerosene pressure lamps.

In Germany, farmers have begun making and using biogas because it makes economic sense to do so. They use the farm's septic tank to create the gas, which they burn in an adapted car engine to produce heat and electricity for the farm.

Biogas is a mixture of methane and carbon dioxide. It is produced by the action of bacteria on vegetable/organic material in airless conditions, which is why the process is also known as 'anaerobic digestion'. The bacteria slowly digest the material (usually animal dung, human wastes and crop residues) and produce a gas which is roughly 60 per cent methane and 40 per cent carbon dioxide.

Producing biogas

When a cow eats grass, bacteria living in the cow's gut assist the digestion process by breaking the food down into simpler chemicals and gases. One of these is biogas. Bacteria excreted with the cow's dung can continue to turn the chemicals in any organic material into gas if the conditions are similar to those in the cow's gut. Light and air must

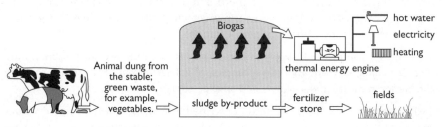

The process of anaerobic digestion

be excluded and the dung has to be kept at a warm temperature, somewhere between 20 and 40°C. The conditions can be met by digging a hole in the ground, which should be lined with brick or stone mortar to stop the dung mixture from leaking out.

Biogas plants

In Nepal, fixed dome plants have been used to collect the biogas produced. A fixed dome plant consists of four basic components: the inlet, the digester, the gas holder and the outlet or expansion chamber.

The inlet, equipped with a mixing device, is used to collect the dung and mix it with an equal amount of water. After the dung and water have been mixed together, a plug at the bottom of the inlet is removed and the mixture, called slurry, flows through a pipe into the digester.

The digester is a flat-bottomed, round chamber, covered with a dome-shaped concrete gas holder. The fixed dome plant needs to be sealed properly in order to prevent any gas leakage. The bacteria thrive on the dung mixture in the digester and create biogas. The gas then rises and is stored in the gas holder before being released into a pipe. The slurry leaves through the outlet or expansion chamber and flows into the compost pit.

Digestion time ranges from a couple of weeks to a couple of months depending on the feedstock and the digestion temperature. The residual slurry is removed at the outlet and can be used as a fertilizer, which increases agricultural production, especially in vegetable growing.

In Germany, about 80 kilometres south of Hamburg, 20 farms are now using animal excrement as a renewable source of energy. Each farmer has a biogas plant which works independently of the other farms, creating an effective decentralized energy system.

The farmers use the heat that is generated from the biogas to warm their homes, the farm buildings and the stables, as well as for hot water. In the summer, the heat is used to dry the harvest. The electricity generated is used in the home – for example, for cooking and lighting – and on the farm, for milking and so on.

Dieter Prenzler is one of the farmers who fuels his farm and buildings by using biogas. The gas is generated from the bodily wastes of 1200 pigs, 4000 hens and the waste from the farmer's own family. There is sufficient gas produced to heat, light and power the farmhouse and its outbuildings in all but the coldest weather or busiest periods.

The animal excrement needed to make biogas is washed into the system through a false floor by gravity and the pig's urine. The droppings and urine from the pigs fall into the space below the floor, and then through a hole into the septic tank.

Other ingredients can also be added, such as hen droppings, waste from the farmhouse lavatory, and straw from the stables, which adds body to the concoction. Waste fats are brought in from the restaurants in Hamburg and they produce 20 times more gas than the excrement from the pigs. The other benefit is that Farmer Prenzler is paid for using the restaurant waste.

The animal wastes are transformed in the septic tank by microorganisms that produce biogas and a high-quality, low-smell fertilizer. The gas powers a standard car engine to create the heat and electricity for the farm and its buildings. So far, only Ford and Opel engines have been used because they have the necessary metallic composition to cope with the sulphur released by the gas. The car engine needs regular maintenance and servicing and the whole system takes up to an hour a day to maintain.

After the farm and its buildings have been heated by the biogas, and the cooking, etc. has been completed, there is still some power left over which the farmers sell to the electricity companies, and a high-quality fertilizer which does not smell. Because it has lost its potent smell, the fertilizer does not have to be dug into the earth like con-

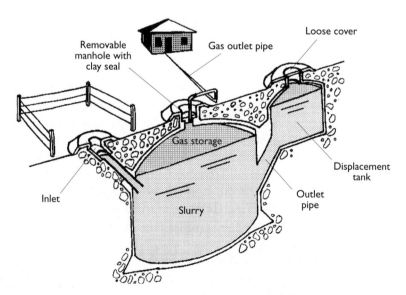

A fixed dome digester
© ITDG

ventional sludge. Instead, it can be sprayed directly on to the crops. Another advantage is that the sludge can be spread at any time of the year.

Outputs

When the conditions are right, one kilogram of dung produces 40 litres of biogas. At full capacity, the biogas stoves use approximately 400 litres of gas per hour, but the user can reduce the size of the flame with the gas tap. Studies show that the average family has been saving between 140 and 180 kg of firewood per month by using biogas. Compost can also be obtained from human waste, although it takes about 75 people to produce the same amount of dung as one buffalo!

Applications for biogas

Biogas has a variety of applications. The table below shows some typical applications for one cubic metre of biogas. Small-scale biogas digesters usually provide fuel for domestic lighting and cooking.

Some biogas equivalents

Application	Capacity of 1m^3 biogas
Lighting	One 60–100 watt bulb for six hours
Cooking	Three meals for a family of five or six
Fuel replacement	0.7 kg of petrol
Shaft power	Able to run a 1 hp motor for two hours
Electricity generation	Able to generate 1.25 kWh electricity

Source: adapted from Kristoferson 1991

Costs

In Nepal, a small government grant has helped to encourage people to invest in the plants, as well as the availability of credit from a special scheme devised by the banks. Once a biogas plant has been installed there is little expense because it costs nothing to fuel it. It also requires very little maintenance in order to operate effectively. The by-product can be used as a fertilizer for the produce being grown in the fields. It adds to production and lowers the dependency on chemical fertilizers, thus increasing savings.

Advantages of biogas

The digestion of animal and human waste yields several benefits:

- the production of methane for use as a fuel
- the waste is reduced to slurry which has a high nutrient content and makes an ideal fertilizer; in some cases, this fertilizer is the main product from the digester and the biogas is merely a by-product
- during the digestion process bacteria in the manure are killed, which is a great benefit to environmental health
- an effective power supply from natural and renewable resources.

Further information

Biogas Support Programme
PO Box 1966
Kathmandu
Nepal
Tel: +977 1 521742/534035/524665
Fax: +977 1 524755
E-mail: snvbsp@wlink.com.np

Biogas Support Programme
DevPart Consult – Nepal
GPO Box 5517
Battisputali
Kathmandu
Nepal
Tel: +977 1 476264
E-mail: devpart@dp.mos.com.np

Hans-Hermann Jacobs
Ilhorn 1
29643 Neuenkirchen
Germany
Tel: +49 5195 9870
Fax: +49 5195 9871

Heinrich Weseloh
Vahlzen 10
29643 Neuenkirchen
Germany
Tel: +49 5195 2582
Fax: +49 5195 2570

Capturing coal-bed methane

China is the largest producer of coal in the world, generating about 1.1 billion tonnes of coal every year. It is rich in coal-bed methane and its deposits account for an estimated one-third of the world's total methane, mainly from underground mines which have high emission levels. Methane is released during coal mining operations, but only 110 of the 600 state-run coal mines have systems in place to recover methane and only 40 of these actually use the recovered methane.

Methane is a greenhouse gas that is 20 to 60 times more reactive than carbon dioxide. Atmospheric concentrations of methane are rising, with adverse effects on the ozone layer as well as increased smog formation and global warming. Fortunately, methane can also be a useful energy source if it is captured and used effectively.

Widespread use of coal in China has created severe local air pollution. Coal-bed methane is a remarkably clean fuel when burned, and resources are distributed throughout the country. There are an estimated 30–35 trillion cubic metres of methane reserves in China. About 500 million cubic metres of methane can power a 20 MW power plant for one year. A 20 MW power plant produces enough power to meet the needs of a small city with a population of 20 000 people. The increased use of methane as a fuel would decrease the use of fossil fuels and associated environmental problems.

The Global Environmental Facility

The Global Environmental Facility (GEF) is a joint venture between the United Nations Development Programme (UNDP), the United Nations Environment Programme (UNEP) and the World Bank. GEF is a financial mechanism that provides grants and concessional funds to developing countries for projects and activities designed to protect the global environment in terms of climate change, biological diversity, international waters and depletion of the ozone layer.

The development of methane energy in China began to take substantial steps when UNDP provided technical assistance to the Chinese Government in the 1990s. GEF and China's Ministry of Coal set up a project to reduce atmospheric methane emissions and recover clean-burning methane as a fuel.

Development of coal-bed methane resources in China

The programme was designed to:

- improve the efficiency of the coal mine methane recovery systems
- use the recovered methane for beneficial purposes that would prevent it from escaping into the atmosphere
- improve productivity and safety of China's coal mines
- develop a community of technologists that could advance the industry into full national proportions.

The project succeeded in developing the capabilities in China for sustained coal-bed methane development by providing:

- advanced technologies and techniques for assessing the coal-bed methane resources and to extract and capture the gas more efficiently and effectively than had been common practice in China
- training of technical personnel in the proper use of the technologies
- exposure of technical and government personnel to developments abroad
- exchange of ideas through technical conferences.

The recovery method

As plant material is converted into coal, large quantities of methane-rich gas are generated and stored within the coal. The presence of this gas can cause explosions during underground coal mining. Absorbed methane is liberated as coal is mined and the escaping gas must either be captured in methane recovery systems, or ventilated to the atmosphere in order to prevent hazardous explosions.

It is essential to extract methane before mining because of the danger to miners. Secondary precautions, such as venting fresh air into the mining shaft to ensure that methane levels remain below 1 per cent and that it is safe for humans to move about in, are taken when the actual mining is being done. The methane is removed from the mines by vertical drilling. There are large pockets in the coal bed where the methane is trapped. After the pockets have been located, the methane is extracted before any mining takes place in that area.

Water is used to create pressure and make a fissure in the coal bed. Once the water has been removed, the methane naturally collects in the fissure and can be pumped through pipes to the surface where it is stored in cylindrical storage tanks. This type of drilling collects the methane efficiently, and practically eliminates release of the gas into the atmosphere. It makes the recovery of coal safer and can ultimately increase the volume of methane captured.

Benefits of the project

Coal-bed methane can be a valuable energy resource that is important to economic development and, if properly captured and used, can result in significant environmental benefits. The production and sale of coal-bed methane can be a source of significant new revenues as well as a means of reducing coal mining costs.

Coal-bed methane has been commercialized in a few cities in China. Pipelines have been installed to connect the fuel to the residential districts and enterprises, which generates considerable revenue to the coal companies. As a result of the increased use of coal-bed methane at Teifa City for cooking in 20 000 homes in place of coal, there have been significant reductions in sulphur dioxide, nitrogen oxides, particulate matter, carbon dioxide and carcinogens in the atmosphere.

In addition to the extra profits received by selling the gas, considerable cost savings in the coal mining operations have resulted from coal-bed methane recovery. For example, the use of the gas substituted coal that would have been burned. Savings were also made from the reduced electric power needs of the mines for ventilating the methane, for safety considerations, by approximately 20 per cent. As a result of reduced electricity requirements for methane ventilation, emissions from the power plant servicing the coal-bed methane have been reduced as well.

Since coal-bed methane production began under the UNDP project, no coal mine explosions have occurred because of the significantly reduced gas levels in the mines. The recovery of methane has substantially reduced the greenhouse gas emissions while improving coal mine production and safety, and local and global air quality.

Further information

United Nations Development Programme
Mahenau Agha
Information Officer
One United Nations Plaza
New York
NY 10017 USA
Tel: +1 212 906 6112
Fax: +1 212 906 6998
E-mail: mahenau.agha@undp.org
Website: www.undp.org/gef

GEF Headquarters
1818 H Street NW
Washington DC 20433
USA
Tel: +1 202 473 0508
Fax: +1 202 522 3240 / 522 3245
Website: www.gefweb.org

Other useful contact:
jdliu@public.bta.net.cn

Other useful website:
http://energy.usgs.gov/factsheets/coalbed/coalmeth

Micro-hydro power

The rural population in Peru is eight million and it is spread over an area of more than 1.2 million square kilometres. The majority of Peru's rural population lives in the Andean highland region. Communities and settlements are very small and in remote locations. The towns and cities in Peru have electricity, but the communities living in the 'cut-off' areas in the mountains have few facilities and little prospect of gaining access to them. Before other services can be introduced to meet basic needs, it is fundamental that power is available to these communities. The cost of expanding the grid of electricity into the widely dispersed population of the mountains is very high and therefore unlikely to happen even in the long term. This has meant that government programmes to develop education, sanitation, transport and health services usually get through to only larger and more accessible settlements.

Energy needs in Peru

In 1996, only 4 per cent of Peru's rural population had access to electricity.

- For light in the home, kerosene and candles are widely used and some slightly better-off families use batteries.
- For cooking, residues from processed crops and wood from the few trees in the area are used. Wood collection is time-consuming and the task is often done by women.
- There are a few water mills in the region for turning grain into flour. These do not generate much power and are not fast enough to meet today's demands. Poor farming families spend a lot of time milling by hand, using animal power, or travelling to other settlements that have a motor-powered mill, for which there is an expensive charge.

Micro-hydro as an energy option

Peru is rich in many energy resources, such as water, petrol, coal and natural gas, all of which are yet to be fully exploited. The potential power to be gained from the country's water resources is estimated at 75 000 MW, which would be enough to meet the energy needs of 20 million people in a Northern country where consumption is much higher per person. Micro-hydro schemes offer a sustainable method of harnessing energy from running water and can be used to generate electricity or drive machinery.

The difference between micro-hydro and larger hydro schemes is the amount of electricity they can produce and the amount of interference with the natural environment needed to produce this energy.

Unlike large-scale hydroelectric schemes, micro-hydro schemes are 'run-of-river' and rarely require the use of a dam for water storage, or artificial lakes to hold the massive volumes of water needed for the larger hydro schemes to operate. Micro-hydro schemes do not interfere with river flow and are not harmful to the environment because they avoid the negative environmental and social impacts associated with projects of large-scale hydro. With micro-hydro, a maximum capacity for the system would be up to 300 kilowatts, which would be a 'stand-alone' system; that is, not connected to the national grid.

Peru has used micro-hydro systems since the beginning of the twentieth century, mostly for use in mining or on large farms. The technology is becoming more accessible to poorer communities because there are small workshops that make micro-hydro turbines and once installed these can be run by communities. Micro-hydro is a viable energy option technically, economically and socially.

Operating a micro-hydro system

A micro-hydro scheme starts operating at the intake weir, where water is diverted from the river. The function of the intake weir is to maintain the water level so that a continuous flow of water is achieved. The intake is usually protected by a trash rack of metal bars, which filters out debris. The water then passes through the settling tank or forebay, which slows the water down so that any suspended particles can settle at the bottom.

The height that the water falls through is known as the 'head'. In a medium or high head installation, water is conducted to the forebay by either a channel or a small canal. In a low head installation, water enters the turbine almost directly from the intake weir. From the forebay tank, water reaches the turbine from a pressure pipe or penstock. PVC piping can be used for this component as it is strong and smooth, which reduces the friction. It is also relatively cheap and easy to transport and install.

The water then meets the turbine, which is located in the powerhouse. The amount of power the turbine is able to produce depends on the distance of the fall, the speed of the flow, and the number of litres per second flowing through the system.

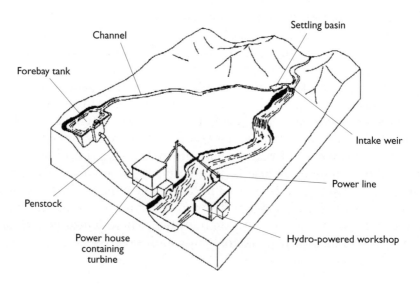

Components of a typical micro-hydro system

Turbines

Turbines come under three broad groups related to head levels – high, medium and low – and then generally into two categories, impulse and reaction turbines.

A micro-hydro system in Peru
© ITDG/Colin Palmer

For medium to high head applications an impulse turbine is usually used. The pressurized water is converted into a high-speed jet by passing through a nozzle. The jet of water strikes the specially shaped buckets or blades of the turbine rotor, which then rotates.

A reaction turbine works in a very different way in that it runs filled with water. The flow of water through the turbine creates different levels of pressure across the blades, which causes them to rotate.

The turbine will sometimes need to be shut down for maintenance so all installations should have a sluice gate or valve at the top of the penstock to close off the water supply. When it is closed, the water is diverted back to the river down a spillway.

Battery power from micro-hydro

Part of the versatility of micro-hydro power is that the delivery of energy from the system can be adapted to suit the needs of a community and modified to get beyond any physical constraints. An ITDG micro-hydro project in Huacataz, Peru, found that the community demand for electricity was for domestic use (lighting and television) and for powering the grain mill. The conventional method of delivering energy is through a centralized scheme that distributes electricity to domestic connections. This was not viable as the population in this community was too widely dispersed. The solution was to introduce a scheme where the grain mill was positioned centrally and served a secondary function as a battery charger. The small charge for the service of milling and battery charging meant that the scheme also raised funds for further community development and covered the running costs of the scheme.

Considerations in installing micro-hydro

Pre-feasibility studies need to be done in the area where the installation of a micro-hydro scheme is being considered, to determine several points:

- A suitable geographic location for a micro-hydro scheme is in an area where there are steep rivers that flow all year round. Areas that give the highest head options should be considered first as they will usually cost less per installed kilowatt. To determine the power potential of the water flowing in a river or stream, it is necessary to determine both the flow rate of the water and the head through which the water can be made to fall. The flow rate is the quantity of water flowing past a point in a given time. Knowing the flow rate will help to determine the appropriate type and size of a turbine.
- The demand for electricity needs to be determined. This will help to decide what size system will be needed; whether or not electricity demands will fluctuate seasonally or remain constant; and whether the demand is domestic or industrial, or a combination of both.

Cost

The initial investment is high but running costs after installing the micro-hydro are low because the water is free. A micro-hydro unit should last for 20 years provided that regular maintenance checks and servicing are carried out. Schemes that provide only power for driving equipment are cheaper to install.

Further information

The Micro-Hydro Programme
ITDG Peru
Casilla Postal 18-0620
Lima 18
Peru
Tel: +51 1 4467324/4447055/4475127
Fax: +51 1 4466621
E-mail: postmaster@itdg.org.pe
Website: http://www.itdg.org.pe

POWER WITHOUT DESTRUCTION   CHAPTER ONE   **15**

Water current turbines

In the jungle areas of Peru most of the rural population (about 1.5 million people) are settled along the banks of the largest rivers, such as the Amazon, Maranon, Ucayali and Napo, because the rivers are the natural means of transport and communication, as well as a source of food. In fact in the Peruvian jungle there are hundreds of small communities that live mainly settled on the river banks, where the river is one of the most important means of life for them.

Energy use in small and isolated communities

Isolated communities or villages like those along the river banks in the Peruvian jungle require energy to satisfy essential needs, such as lighting, vaccine preservation through refrigeration, and communications equipment. Small kerosene lamps or burning wood are used for lighting. Those villages that are closer to towns such as Iquitos and Pucallpa have found an appropriate way of providing lighting and power for radios by using vehicle batteries. These batteries have to be recharged regularly in towns, which involves an expensive journey that could take a whole day, plus the charging cost.

Unfortunately, the usual sources of energy are not readily available in this area. Conventional hydro power generation is not practical due to the topography; diesel engines require fuel, oil, spare parts and specialized maintenance; solar energy is still expensive and the solar radiation is weak because of cloud cover; and wind speeds are too low to generate power. The one energy source that is present in abundance is the rivers themselves which, in most cases, have speeds high enough for power generation.

In this situation, river current turbines become a very interesting option which should be tested and promoted strongly in order to benefit as many settlements as possible. If the introduction of small units proves to be successful, it is possible to extend the range of power capacity not only by adding units but also by designing and building larger machines.

Water currents

The idea of using the kinetic energy in river and canal currents to generate energy started in the late 1970s. A current speed of 1m/s represents an energy density of 500 watts per square metre of cross-section. Water current turbines extract some of this energy and convert it into electricity or mechanical shaft power to drive a pump. These turbines are quiet and non-polluting in use.

The energy available at a site is dependent on the speed of the water and its depth. Since water speed and depth can change throughout the year, accurate knowledge of any potential site is essential. It has also been found that river speeds can vary considerably over short distances and so site surveys are necessary to ensure the best performance from the machines. Accurate measurement of water speed is particularly important as the energy available is proportional to the cube of the speed; for example, a 10 per cent increase in water speed from 1 m/s to 1.1 m/s will give a 33 per cent increase in energy density available.

Battery-charging water current turbines

Two companies – Marlec Engineering, leading wind charger manufacturers, and Thropton Energy Services, water turbine specialists – have joined forces to develop a battery-charging water current turbine, which they call the Amazon Aquacharger.

The 1.8 m diameter turbine has been designed to generate free power when moored on any river or canal deeper than 1.75 m and flowing at between 0.45 m/s (1 mph or 0.87 knots) and 1.5 m/s. The turbine can be mounted on the back of an ordinary open boat, providing it is more than 5 metres in length. It starts charging batteries at 0.5 m/s, can charge up to six 12- or 24-volt batteries simultaneously and will operate continuously for 24 hours per day as needed, provided the water speed stays below 1.5 m/s. At water speeds above

Battery-charging water current turbine
© ITDG/Saul Ramirez

1.5 m/s the mechanical loading on the turbine becomes excessive and so the turbine will automatically shut down to avoid damage. The exceptionally low water current speed required to start the battery-charging generator has been achieved through aerodynamic blade design and the use of a high-efficiency low-friction alternator.

Power from the turbine is stored in the batteries and between ten and 20 car batteries can be recharged every day at a good site. This power is then available for running 12 V appliances such as lighting, vaccine refrigerators, portable televisions, radio communications equipment, and inverters to operate 240 V AC appliances. Larger machines of a similar design can also be supplied for electricity supply or irrigation.

Specifications of the water current turbine

The water current turbine uses a large, three-bladed rotor which is powered by the kinetic energy in a free stream of water. The turbine is installed on a floating base, which can be in the form of a pontoon, an ordinary open boat or any other similar structure. This structure is tied to the bank by means of a mooring post. The fixing of this post is the only civil engineering work needed to install the turbine. A rigid walkway connects the floating base to the riverbank. This walkway has three functions – it gives access to the machine; supports and protects power cables; and stops the turbine from drifting towards the bank. The shaft of the turbine is coupled either to an electrical generator or to a small pump if it is to be used to raise water on to the riverbank for irrigation. Only the rotor and one bearing are below the waterline. The transmission system and generator are all installed above the water on the floating base. This configuration prolongs the life of the machine and makes access to the components easier. To keep the capital cost down the turbines have been designed for local manufacture.

The Aquachargers used in Peru are stand-alone units and have a maximum power output of about 2 kW, depending on the water resource available. Larger power capacities can be obtained by installing two or more units in parallel. The amount of energy available in the flowing water depends on the area swept by the turbine as well as the cube of the water speed. For example, on the Nile in Sudan, Garman turbines (developed by Thropton Energy Services) with a swept area of 10 m^2 operating in a river speed of 1 m/s generate about 3 kW of power per unit. In places where the river speed is higher, the energy available increases greatly and, with 1.2 m/s flow, the same machine can produce up to 5 kW of power. It should be noted that this size of turbine needs a site with a water depth of at least 4 m.

Safety measures

The system incorporates a furling device that lifts the turbine out of the river if the water current speed exceeds the preset maximum. As the unit furls, the turbine is electrically braked to avoid the blades freewheeling, reducing unnecessary wear and tear. Should sudden flood conditions occur (water speeds above 1.5 m/s), the automatic furling system will activate. If the mainstream water speed is too fast, the turbine can easily be moved closer to the bank and manually reset. This is achieved by two people carrying the bank-end support of the rigid walkway away from the river's edge. The turbine will then automatically float towards the bank. The rotor is also protected against striking the riverbed or floating debris.

Should river conditions change permanently and make the site unsuitable, the turbine can be furled manually and the whole unit can be easily dismantled to fit on to a small truck and transported to alternative sites with the minimum of labour and effort.

The charge controller included senses the battery terminal voltage and automatically stops the turbine rotation when full charge is achieved; it continues to monitor the battery voltage state and restarts the turbine as the charge level decreases.

Advantages of the battery-charging water current turbine

This cost-effective solution is ideal for rural communities, health centres, schools, missions, battery-charging stations, tourist lodges, border posts, radio communication sites or research stations, and offers the following benefits:

- affordable power to isolated areas
- easily transportable so many users can benefit
- completely independent operation
- continuous power generated 24 hours/day
- non-polluting
- simple to install, operate and maintain.

Further information

Saul Ramirez
Programa de Energía ITDG-Peru
Av. Jorge Chávez 275
Lima 18
Peru
Tel: +511 447 5127 / 444 7055
Fax: +511 446 6621
E-mail: saul@itdg.org.pe
Website: www.itdg.org.pe

Marlec Engineering Company Ltd
Rutland House
Trevithick Rd
Corby
Northants NN17 5XY
United Kingdom
Tel: +44 (0)1536 201588
Fax: +44 (0)1536 400211
Email: sales@marlec.co.uk

Dr B. Sexon
Thropton Energy Services
Physic Lane
Thropton
Northumberland NE65 7HU
United Kingdom
Tel: +44 (0) 1669 621288
Fax: +44 (0) 1669 621288
E-mail: throptonenergy@compuserve.com
Website:www.ourworld.compuserve.com/homepages/throptonenergy/homepage.htm

Harnessing wave energy

Fossil fuels such as coal and oil are not renewable over the span of human generations, and their use is subject to increasing environmental concerns as it contributes to global warming and acid rain. To meet the growing energy needs in the UK, engineers in coming decades will be challenged to generate power economically from renewable energy sources. Despite the fact that nearly 75 per cent of the Earth's surface is covered with water, waves are a largely unexplored source of energy, compared with the progress that has been made in harnessing the sun and wind.

It has long been recognized that near-shore wind and wave resources offer outstanding potential for generating renewable power, and it is estimated that in the UK alone the recoverable energy resource exceeds electricity demand.

Wave power applications

- electrical power generation
- coastal protection
- harbour development
- satellite oilfield power
- aquaculture development
- desalination.

Waves are a free and sustainable energy resource created as wind blows over the ocean surface. They are a store for solar and wind energy. The greater the distances involved, the higher and longer the waves will be. Energy is stored in this way until it reaches the shallows and beaches of our coasts where it is released, sometimes with destructive effects.

Until recently, the commercial use of wave power has been limited to small systems of tens to hundreds of watts aboard navigation buoys. As the buoy heaves up and down in waves, the oscillating water column in the centre pipe of the buoy's hull acts like a piston, alternately pushing air out the top of the pipe and drawing it in. This pneumatic power can be converted directly to sound through a foghorn, or indirectly to light by spinning a turbogenerator, which charges an electrical storage battery.

Wave power modules are built to harness energy from waves and transform it into electrical power. By absorbing the incoming energy, power modules create an area of calm water behind them, contributing toward coastal defences and producing a valuable area for other commercial and recreational marine activities. This protected area can be used to create self-financing harbours and breakwaters. Their installation can bring positive environmental and economic spin-offs, such as protection of threatened areas of coastline or provision of an environment suitable for aquaculture development.

Artificial reefs substantially improve the local marine bio-density, attracting shoals of fish and providing habitats for colonization by commercially valuable species. Wave power modules can act as these artificial reefs and also have the potential to improve the local inshore marine harvest. Benefits will be greater in areas presently sparse in marine life and devoid of suitable substrate for settlement.

Shortage of drinking water is a cause of much suffering and a major limitation to agricultural development in many parts of the world. Wave power devices can provide supplies of potable fresh water through on-board seawater desalination to help alleviate these problems. This is of particular value in dry coastal areas with strong wave regimes.

The regular and powerful wave climates found in many locations where water shortages exist result in ideal locations for the generation of power from waves. Many of these peripheral locations currently rely on diesel generators to provide electricity to power their desalination plants. It is in these regions in particular that the benefits of wave power are considerably magnified.

Wave power technologies

Established in 1992 by Professor Alan Wells FRS, the inventor of the Wells turbine, a British company called Wavegen has developed a range of energy modules for clients to exploit the unlimited wave energy resources in the shoreline, near-shore and offshore environments.

Wavegen modules capture waves, converting the stored energy back into pressurized air which drives a turbine. The turbine rotates a generator, transforming this movement into electrical energy. In operation, the air-driven turbogenerator is the only rotating component, reducing plant maintenance to a minimum.

This technology leads the way in harnessing the vast untapped resources of the world's oceans. The units are capable of generating electricity at a price that is competitive with fossil-fuelled power stations and other renewable energy sources. Furthermore, the modules generate power that is emission free.

Wavegen's power devices comprise two elements – a collector to

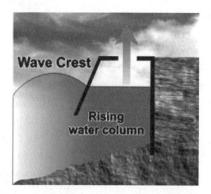 

A pneumatic Wells turbine collects energy from an artificial blowhole
Source: Wavegen

capture the wave energy and a turbogenerator to transform the wave power into electricity.

Oscillating water column

Oscillating water column devices use a pneumatic Wells turbine to collect energy from an artificial 'blowhole'. In this design, a large structure is built, usually on or near the shore, containing a chamber with an opening to the sea below the waterline, through which sea water is free to enter and leave. As waves strike the device, the water level within the chamber rises and falls, pushing air out through an opening at the top. This column of air, contained above the water level, is alternately compressed and decompressed by the movement to generate an alternating stream of high velocity air in an exit blowhole. If the air stream is allowed to flow to and from the atmosphere via a pneumatic turbine, energy can be extracted from the system and used to generate electricity.

Power take-off – the turbogenerator

Wells turbines are used to power the electricity generators. These turbines have the unique property of rotating in the same direction regardless of the direction of fluid flow around the turbine blades. This design allows the turbine to spin both as air is expelled from the chamber by the rising water and as it rushes back in when the wave subsides. Thus, the turbines continue turning on both the rise and fall of wave levels within the collector chamber. The turbine drives the generator, which converts this power into electricity.

The wave energy converter

Wave energy increases with distance from the shore. The Pelamis wave energy converter is a device that has been designed by another British company, Ocean Power Delivery. It is a semi-submerged, articulated structure composed of cylindrical sections linked by hinged joints. The wave-induced motion of these joints is resisted by hydraulic rams which pump high-pressure oil through hydraulic motors via smoothing accumulators. The hydraulic motors drive electrical generators to produce electricity. Power from all the joints is fed down a single umbilical cable to a junction on the seabed. Several devices can be connected together and linked to shore through a seabed cable. A 750 kW device will measure 150 metres in length and 3.5 metres in diameter.

Benefits of wave power

- Oceans cover three-quarters of the earth's surface and represent a vast natural energy resource in the form of waves.
- The World Energy Council estimates that energy equivalent to twice the world's electricity production could be harvested from the world's oceans.
- Waves are a renewable source of power which is emission free.

The articulated joints of the Pelamis converter are moved by the waves
Source: Ocean Power Delivery Ltd

Further information

Wavegen
50 Seafield Road
Longman Industrial Estate
Inverness IV1 1LZ
United Kingdom
Tel: +44 (0) 1463 238094
Fax: +44 (0) 1463 238096
E-mail: enquiries@wavegen.co.uk
Website: www.wavegen.co.uk

Ocean Power Delivery Ltd
2 Commercial Street
Edinburgh EH6 6JA
Scotland
United Kingdom
Tel: +44 (0) 131 554 8444
Fax: +44 (0) 131 554 8544
E-mail: enquiries@oceanpd.com
Website: www.oceanpd.com

Wind power

Wind energy offers the potential to generate substantial amounts of electricity without the pollution problems of most conventional forms of electricity generation. The equipment used for wind energy technology is being continually improved to make it cheaper and more reliable. It is therefore expected that wind energy will become even more economically competitive over the coming decades.

Wind generation for developing countries

Unlike the trend towards large-scale grid-connected wind turbines seen in the North, the more immediate demand for rural energy supply in the South is in the 5–100 kW range. These smaller machines can be connected to localized micro-grid systems and used in conjunction with diesel generating sets and/or solar photovoltaic systems. Currently, the use of wind power for electricity production in developing countries is limited, the main area of growth being for very small battery-charging wind turbines (50–500 watt). Other applications for small wind machines include water pumping, telecommunications power supply and irrigation.

Power in the wind

Wind is simply the movement of air from one place to another. There are localized wind patterns caused by temperature differences between land and seas, or mountains and valleys. Wind speed generally increases with height above ground because the roughness of ground features, such as vegetation and houses, causes the wind to slow down.

For areas with mean annual wind speeds of above 4–5 metres per second, wind-powered electricity generation is an attractive option. Wind speed data can be obtained from wind maps or from the meteorology office. Unfortunately, the general availability and reliability of wind speed data is extremely poor in many regions of the world. It is important to obtain accurate wind speed data for the site under consideration before any decision can be made as to its suitability.

The power in the wind is proportional to:

- the square of the rotor diameter
- the cube of the wind speed
- the air density, which varies with altitude.

The actual power that can be used will depend on several factors, such as the type of generator and rotor used, the sophistication of blade design, friction losses, and losses in the pump or other equipment connected to the wind machine. There are also physical limits to the amount of power that can be extracted from the wind. Any windmill can extract only 59.3 per cent of the power from the wind (this is known as the Betz limit) and, in reality, the maximum is usually around 45 per cent for a large electricity-producing turbine and around 30–40 per cent for a windpump.

The component parts of a wind turbine

The rotor and blades

Most electricity-producing turbines have three rotor blades, although some have two, which are usually on a horizontal axis connected to the hub. Rotor diameters can be up to 65 metres, while smaller machines have rotor diameters of around 30 metres. The longer the blades, the larger the area swept by the rotor and hence the greater the energy output. Blades are made of glass-fibre-reinforced polyester or wood-epoxy. The shape of the turbine blade is designed so that, when air passes over it, the rotor to which it is attached turns.

The turbine head or 'nacelle'

The rotor is linked by a shaft direct to the nacelle, which contains a gearbox and a generator. As the blades rotate, the shaft is turned to drive the generator and produce electricity. Most machines have gearboxes, although the number with direct drive is increasing.

The yaw mechanism

The wind direction is detected by sensors which control the yaw mechanism. This turns the turbine so that it always lines up with the wind.

The tower

Towers are mostly tubular and made of steel. Towers range from 25 to 80 metres high and act as a support to the nacelle and rotor.

Electricity produced by the generator comes down the cables in the tower and passes through a transformer into the electricity network.

Large turbines are built on a concrete foundation.

When a turbine comes to the end of its working life, it is easy to dismantle and its scrap value will cover the cost of dismantling. The base can be dug up or covered, leaving little trace behind.

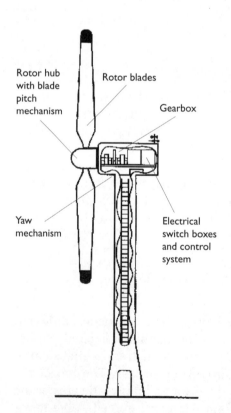

The parts of a wind turbine

Preventing damage in high winds

Most machines operate at a constant speed of 15–50 revolutions per minute, though new designs have variable speed. Power is controlled automatically as wind speed changes. When wind speed becomes too high, the turbine is shut down to avoid damage.

Stall and pitch control

The most common design of wind turbine is the three-bladed, stall-controlled, constant-speed machine. Stall and pitch control are the means of controlling power output and describe the design of the blades. On a pitch-controlled machine, the angle of the blades can be actively adjusted by a machine control system. This is similar to having brakes on the blades because when they are fully 'feathered' they will be stationary in the wind.

Stall control is sometimes known as 'passive control' because the blades are aerodynamically designed to perform the same function as a pitch control mechanism, without the moving parts. The twist and varying thickness of the blade mean that when wind speed becomes too high, turbulence occurs behind the blade, shedding some of the wind's energy and minimizing power output at high wind speed. The tips of the blades also have brakes so that they can be brought to a complete standstill if necessary.

Principles of wind energy conversion

There are two primary physical principles by which energy can be extracted from the wind. These are through the creation of a lift or drag force (or through a combination of the two).

The basic features that characterize lift and drag are:

- drag is in the direction of airflow
- lift is perpendicular to the direction of airflow
- generation of lift always causes a certain amount of drag to be developed
- with a good aerofoil, the lift produced can be more than 30 times greater than the drag
- lift devices are generally more efficient than drag devices.

The horizontal axis wind generator, as shown in the diagram, is the most commonly used type of wind machine and uses lift forces to harness the wind. For electricity generation, a small number of blades will give the highest rotational speed needed for most generators.

Small wind generators

A typical small wind generator has a rotor that is directly coupled to a generator which produces electricity either at 120/240 volts AC for direct domestic use or at 12/24 volts DC for battery charging. A tail vane keeps the rotor orientated into the wind. Some wind machines have a tail vane which is designed for automatic furling (turning the machine out of the wind) at high wind speeds to prevent damage. Larger machines have pitch-controlled blades (that is, the angle at which the blades meet the wind is controlled) which achieve the same function. The tower will often be guyed to give it support.

Grid-connected or battery charging

Depending on the circumstances, the distribution of electricity from a wind machine can be carried out in one of various ways. Commonly, larger machines are connected to a grid distribution network. This can be the main national network, in which case (providing an agreement can be made between the producer and the grid) electricity can be sold to the electricity utility when an excess is produced and purchased when the wind is low. Using the national grid helps to provide flexibility to the system and dispenses with the need for a back-up system when wind speeds are low.

Micro-grids distribute electricity to smaller areas, typically a village or town. When wind is used for supplying electricity to such a grid, a diesel generator set is often used as a back-up for the periods when wind speeds are low. Alternatively, electricity storage can be used but this can be a more expensive option. Hybrid systems use a combination of two or more energy sources to provide electricity in all weather conditions. The capital cost for such a system is high but subsequent running costs will be low compared with a pure diesel system.

In areas where households are widely dispersed or where grid costs are prohibitively high, battery charging is an option. For people in rural areas a few tens of watts are sufficient for lighting and a

Wind turbines are particularly useful in windy rural areas with no electricity grid connection
© Hugh Piggott

source of power for a radio or television. Batteries can be returned to the charging station occasionally for recharging. This reduces the inconvenience of an intermittent supply due to fluctuating wind speeds. There are 12 and 24 volt DC wind generators available commercially which are suitable for battery-charging applications. Small wind generators of this size (50–500 watt) can also be used for individual household connection.

Economics

The cost of producing electricity from the wind is heavily dependent on the local wind regime. The power output from the wind machine is proportional to the cube of the wind speed and so a slight increase in wind speed will mean a significant increase in power and a subsequent reduction in costs. For example, if the wind speed doubles, the power in the wind increases by a factor of eight. It is therefore worthwhile finding a site that has a relatively high mean wind speed. It is also worth bearing in mind that a wind machine will operate at its maximum efficiency for only some of the time it is running, because of variations in wind speed.

Capital costs for wind power can be high, but running costs are low and so access to initial funds, subsidies or low-interest loans are an obvious advantage when considering a wind–electric system. A careful matching of the load and energy supply options should be made to maximize the use of the power from the wind – a load that accepts a variable input is ideally matched to the intermittent nature of wind power.

Environmental concerns

Wind power is clean and renewable. There are, however, some environmental considerations to keep in mind when planning a wind power scheme.

- Noise – wind rotors, gearboxes and generators create acoustic noise when functioning; this needs to be considered when siting a machine.
- Visual impact – modern wind machines are large objects and have a significant visual impact on their surroundings. Some argue that it is a positive visual impact, others to the contrary.
- Electromagnetic interference – some television frequency bands are susceptible to interference from wind generators.

Local manufacture

Small and medium-sized machines can be produced locally, and are much cheaper than imported machines. This enables manufacturers to make minor modifications during the production process, in order to match systems with desired end-uses and with the local conditions.

The rotor blades can be made locally from laminated wood, steel, plastics or combinations of these materials, depending on what is available, while some of the machinery components can be made by small engineering workshops. Other parts, including special bearings, gearboxes, generators and other electrical equipment, may have to be imported if they are not available in the country of assembly. Towers can be made of welded steel, preferably galvanized, which can usually be manufactured in local engineering works, while the foundations can be cast from reinforced concrete on site.

Further information

Intermediate Technology Development Group
Schumacher Centre for Technology and Development
Bourton Hall
Bourton-on-Dunsmore
Rugby
Warwickshire CV23 9QZ
United Kingdom
Tel: +44 (0)1926 634400
Fax: +44 (0)1926 634401
E-mail: itdg@itdg.org.uk
Website: www.itdg.org

Other useful websites:
www.bwea.com
www.windpower.dk
www.cranfield.ac.uk/sme/ppa/wind/
http://homepages.enterprise.net/hugh0piggott/sri_lanka/

Windpumps

Water pumping is one of the most basic and widespread energy needs in The Philippines, as in most of the rural areas of the world. It has been estimated that half the world's population does not have access to clean water supplies. Sometimes people have to collect water two or three times a day and each time it takes two or three hours. A windpump gives improved access to water supplies and is an affordable technology that can provide the answer to water shortages and droughts.

The Philippine Government aims to use renewable energy sources to achieve energy self-sufficiency by the year 2005. The Philippines are strategically located on the Asia Pacific monsoon belt, which gives the 7100 islands high potential for using wind energy. Throughout the year there is an almost constant supply of wind. The south-west monsoon wind blows from May until late September and the north-east monsoon starts early in October and blows until April. The average recorded wind velocity in The Philippines ranges from 1 to 8 metres per second.

Water pumping is one of the most widespread energy needs

Wind energy is currently used in The Philippines for mechanical water pumping and for small-scale electricity generation. Mechanical windpumps are used for domestic water supply for households and apartments; vegetable gardens and supplementary irrigation; water supply for poultry and livestock; and water supply for petrol stations.

Windpump designs

Windpumps can be produced with basic workshop skills and labour. In The Philippines they are manufactured locally using local and imported designs. Most of them are second-generation types (low-cost and low start-up wind speed) with multiple twisted blades. They can be made from locally available materials.

The windpump is made of 24 or 33 twisted blades and can have a rotor diameter ranging from 1.5 to 7.5 metres. It is equipped with a tail vane that directs the rotor perpendicular to the wind direction and a spring-loaded side vane which serves as a safety device for the windpump in case of high wind. The rotor, side vane and tail vane are bolted to the head assembly, which is elevated by a three- or four-legged lattice tower to a height of 6 to 10 metres, depending on the location where the machine is to be installed.

Mechanical windpumps can pump water at a rate of 5 to 160 cubic metres per second. The latest designs have a start-up wind speed of 1.5 m/s and will shut off automatically at 7 m/s.

In the Philippines project, it takes only 36 hours and four men to erect a windpump. The blades start generating energy with wind speeds from 7.5 kilometres per hour. The optimum speed is 25 kilometres per hour. The blades power a shaft which creates energy for the water to be pumped from the water table. The pumps require minimum maintenance.

Wind machines are highly dependent on wind, and the selection of the type of windpump, model and size will depend on the site. The quantity and the depth of the source of water also need to be considered. As with any technology drawing water from the water table, if the area is overpumped, the water table drops and sea water can seep in. The ideal location for a windpump is therefore in close proximity to a river.

Consideration also needs to be given as to whether the windpump is intended for household use, where continuous operation is necessary at a lower output, or for agricultural use, where large volumes of water are necessary but for shorter periods.

The wind resource potential, the wind speed, the diameter of the rotor, the volume of the water to be pumped, and the suction and discharge head will all affect the economics of a windpump.

Irrigation systems

Agriculture, which is the primary industry of The Philippines, is one of the areas where wind energy use can be addressed. The introduction of high-yielding crop varieties and improved cultural practices, intensive production of high-value or key crops, introduction of new breeds of animals, and the establishment of agro-industrial processing plants all necessitate alternative sources of energy to meet the water requirement and to supply low-cost electricity.

Wind is a high potential source of energy and if used for irrigation purposes would mean a reduction in the cost of production and therefore increased competitiveness in the world market. Using windpumps for irrigation purposes can often be more economical than fossil fuels. Furthermore, water could be supplied to crops at any time throughout their growing period.

For surface irrigation systems, the windpump is directly coupled to a small farm reservoir or to a concrete tank for irrigating rice and vegetable crops. Water pumped by the machine is stored in a reservoir or tank to provide the amount of water needed for land preparation and crop maintenance.

For a pressurized irrigation system, the windpump is coupled to a header tank where water lifted from the well is stored to operate drip or sprinkler systems. From the tank, a series of drippers can be tapped to the lateral lines to supply water to fruit trees and other crops. Micro-sprinklers can also be used for this system to supply water to cut flowers or nursery seedlings in greenhouses.

A high performance windpump in Kenya

Cost

As a natural source of energy, the wind is freely available and produces none of the harmful pollutants associated with other fuels. Unlike coal, oil and gas, it is constantly renewable and will never run out. Wind energy can also be one of the most economical ways of producing power in small quanitities.

The capital costs for wind machines can vary considerably, depending on the performance required from the system, which affects the choice of design. The cost of materials and labour, for both manufacturing and installation, and the cost of shipping can also be significant.

In the Philippines project, with capital costs from as little as US$150, the windpump technology is now cheaper than paying for either water or the electricity needed to pump it from the ground.

Although windpumps can require a relatively high initial capital investment, the total cost over the life of the pumping system is about half of that for a conventional motorized water pumping system.

With a relatively small amount of maintenance, wind machines can be expected to have a life span of about 10 to 20 years – double that of motor-driven pumps – and there are no fuel costs over the whole of this period. Motorized pumpsets require frequent maintenance in the form of oil changes, engine overhauls, replacement parts and checks on wiring, etc., all of which need skilled, specialist labour which might be expensive or simply not available. During the whole lifetime of the motorized pumpset, it will be consuming energy in the form of electricity or fossil fuel.

Economics show that mechanical windpumps are highly competitive with electric and diesel fuel pumps in terms of the overall cost of water pumped per cubic metre. This is especially true in areas that are inaccessible to the electric grid and where the transportation of fuel increases the running costs of the system.

Advantages of windpumps

- Windpumps eliminate the use of imported fossil fuels, which are generally expensive.
- Windpumps have low maintenance costs.
- Windpumps ensure a year-round supply of water, especially during windy periods.
- Windpumps are environmentally friendly – they do not burn fuel.
- Windpumps can be used for domestic purposes or as irrigation

systems and are suitable for rice, vegetables, fruit trees, nurseries, greenhouses, flower gardens, lawns, etc.
- Prices of fuel oil and electricity are increasing; therefore, by using wind energy for pumping water and for supplying electricity, energy-related expenses are reduced and consequently products become more competitive in the world market.

Further information

Alexis Belonio
Wind Energy Association of The Philippines
c/o CPU-ANEC
College of Engineering
Central Philippine University
5000 Iloilo City
The Philippines

Dan Dorillo
Condor Windpumps
Iloilo City
The Philippines

Solar thermal power

Solar thermal power is a particularly appropriate energy source for countries located in the 'Sunbelt': that is, countries that are within 30 degrees of the equator, where there is high direct solar radiation all year round. Solar radiation is the largest renewable energy resource and has greatest potential in the Sunbelt.

There are two main types of technology for converting energy from the sun into electricity. One is known as solar electricity – photovoltaics – where sunlight is converted directly into electricity via solar cells. This technology is most appropriate for small-scale applications. The other is solar thermal power.

Solar thermal power

The two major technologies for large-scale application of solar thermal power are the parabolic trough and power towers. There is also the parabolic dish system, which has a great deal of development potential because the applications are suitable for remote power supply and smaller energy needs.

Solar thermal power uses different media to create heat. The heat is then used to convert water to steam which will power a conventional steam turbine to produce electricity. Fossil fuel is sometimes used as a back-up to heat the water in the boiler, so that the plant can still produce energy on demand even when the sun is not shining.

The parabolic trough

The parabolic trough or 'solar farm' uses long parallel rows of identical 'concentrator modules', which are glass mirrors in the shape of troughs. The trough shape ensures that energy from the sun is maximized and concentrated to reflect on to the absorber tube.

The trough collector is placed on an axis to allow it to track the sun from east to west. A heat transfer medium, usually oil, is in the absorber tube. The concentrated energy from the mirrors will heat up the oil to temperatures of 400°C. The hot oil is then able to heat the water in the boiler and convert it to steam which will drive a steam turbine to produce electricity.

A typical solar thermal plant is able to provide 2000–2500 full load hours per year.

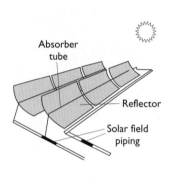

A parabolic trough

The power tower

The principle of harnessing and concentrating sunshine with mirrors onto a receiver is applied using the 'power tower'. In this case, rather than being in rows, the mirrors, referred to as heliostats, are placed in a circular pattern, at the centre of which is the tower. At the top of the tower is the receiver, which contains a fluid, such as water, air, molten metal or liquid salt. The heated fluid from the receiver then goes to a power block, which is used to power a steam turbine.

The power tower is at an earlier stage of development than the trough system, but tests with different heat transfer media show the power tower system is able to produce higher temperatures.

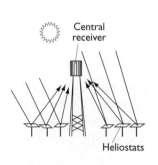

A power tower with heliostats

The benefits of solar thermal technology

The key benefit of solar thermal technology is for the environment. The carbon dioxide emissions from conventional power production alone account for 50 per cent of the harmful gases contributing to the greenhouse effect. The hybrid solar thermal plants in operation in California, where fossil fuel is used as back-up, help to reduce the emissions of carbon dioxide, nitrogen oxide and sulphur dioxide because the plants reduce fossil fuel usage. A typical 80 megawatt solar trough power plant saves 4.7 million tonnes of carbon dioxide emissions from being released into the atmosphere. It also saves about two million tonnes of coal from being used during its 25 years of useful life.

Solar power is sometimes considered to be a land-intensive technology because of the area of land needed for the sites. However, the amount of energy a solar thermal plant produces from a given area is more than would be produced by a large hydroelectric scheme on a similar-sized site. Furthermore, desert land which is otherwise redundant is the most suitable for solar thermal plants and it is unlikely that using such a location will have a negative impact on settlements and habitats.

Diversifying energy resources means there is less reliance on fossil fuels whose prices fluctuate and are subject to increase due to their depleting supplies.

Costs

The initial investment costs of installing a solar thermal power facility are high, with the solar panels or 'solar field' accounting for half of the total cost of the plant. Solar thermal systems in commercial operations are designed to integrate solar power into conventional fossil fuel plants, which can significantly reduce fuel bills as most of the power supply is free from the sunshine.

Future outlook for solar thermal power

Investment to encourage growth in the solar industry is discouraged by the lower cost of fossil fuel plants. Despite this obstacle, conventional coal- or oil-fired steam plants supported by small solar fields are able to compete with large-scale fossil fuel power plants. The strong environmental advantages of integrating solar thermal power to conventional coal-fired plants are cost-efficiency and environmental friendliness.

Financial investment prospects are improving as the international commitment to the environment becomes firmer. Schemes set up by bodies such as the World Bank, UNCED (United Nations Commission on Environmental Development), and the European Union all aim to provide financial incentives and earmark funding for solar thermal projects for developing countries in the Sunbelt.

Developments in parabolic trough technology have led to new methods and designs for integrating the solar technology with gas-fired plants, to give lower investment costs, better conversion efficiency and lower electricity costs.

While power tower technology is still in an earlier stage of development, its prospects for successful commercial growth look increasingly positive. A project in California using liquid salt as a heat transfer medium would be able to replace fossil fuel back-up as thermal energy storage is built into the design to run the plant on demand. Power towers are considered to have good long-term prospects because of the high conversion efficiencies and low electricity costs, particularly in large unit sizes (100–200 megawatts).

Further information

Pilkington Solar International
Mühlengasse 7
D-50667 Köln
Germany
Tel: +49 221 925 970-0
Fax: +49 221 258 11 17

Wood pulp for power

Austria is one of the most densely wooded countries in Europe, with about 45 per cent of its area being covered by forests. Wood, like biologically derived material of any kind, is known as 'biomass'. Biomass is still the main source of energy for more than half the world's population for domestic energy needs. Currently, energy from biomass provides about 13 per cent of Austrian primary energy consumption. About 60 per cent of this energy is used to power traditional stoves and boilers, fired with logs.

Biomass is available in varying quantities throughout the world – from densely forested areas in the temperate and tropical regions, to sparsely vegetated arid regions where collecting wood fuel for household needs is a time-consuming and arduous task. Biomass is also used for non-domestic applications.

Trees, plants, crop residues, animal and human waste, household or industrial residues used for direct combustion to provide heat, are known as 'solid biomass'. Often the solid biomass will undergo physical processing such as cutting, chipping, briquetting, etc. but will still retain its solid form. Crop and industrial biomass residues are widely used to provide centralized electricity production or other commercial end-uses.

Combustion of biomass

Solid biomass needs to undergo combustion in order to be converted into useful heat energy. The combustion process can take place within a basic open fire used for cooking or heating, although other technologies exist for extracting energy and converting it into heat or power for medium and large-scale operations. Combustion efficiency varies according to the fuel and moisture content. The design of the stove or combustion system also affects the overall thermal efficiency.

All biomass contains moisture, and this moisture has to be driven off before combustion can take place. The heat for drying is supplied by radiation from flames and from the stored heat in the body of the stove or furnace. The dry biomass is then heated and, when the temperature reaches between 200 and 350°C, gases are released. These gases mix with oxygen and burn, producing a yellow flame. The heat from the burning gases is used to dry the fresh fuel and release more gases. Once all the gases have been burnt off, charcoal remains. At about 800°C, the charcoal oxidizes or burns.

Biomass use in Austria

The Austrian Government has set up a major action programme to encourage the industrial use of biomass, supporting companies in switching their heating systems to biomass or supplying energy from biomass to third parties.

Sawmills, paper and wood pulp producers, woodworking industries such as construction, furniture and ski manufacture, and craft workshops are the main users of biomass energy, usually as a by-product of their own activities.

Approximately two-thirds of the timber (without bark) that is processed in sawmills is converted to sawn timber. The remaining timber and the bark are used for further processing and for energy production. As most sawmills are equipped with bark-peeling machines, wood residues tend to be bark-free which makes them suitable for further processing in the paper and wood pulp industry, where they fetch a higher price. Therefore, it is mostly the bark that is used for energy production.

Residues of the paper and wood pulp industry, such as bark and black liquor from pulp production, are mainly used to fulfil the industry's own electricity and heat demand. Austria's woodworking industry also produces a considerable quantity of waste wood. About 70 per cent is used as fuel for producing energy and about 30 per cent is processed further, mainly in paper and board production.

Biomass district heating plants

In recent years, a new technology for providing domestic heating in rural areas has been introduced: small-scale district heating plants that use wood chips, industrial wood wastes and straw as fuel. Now more than 300 biomass district heating (BMDH) plants have been established in rural villages in Austria.

A BMDH system incorporates a big furnace fuelled with biomass to heat water that passes through a pipe grid and supplies the energy for the heating of individual houses in a village with between 500 and 4000 inhabitants. Accordingly, the size of BMDH plants varies from a few hundred kilowatts up to 8 MW, with corresponding grids between 100 metres and 20 km. About two-thirds of all the plants have a power of less than 1500 kW.

Biomass district heating has a number of significant advantages compared with traditional heating systems. It substitutes all fuel-handling work, provides continuous heat and reduces emissions of

POWER WITHOUT DESTRUCTION   CHAPTER ONE   41

A biomass district heating system in Ober Zeiring
Source: EVA

predominantly old and technically poor individual heating systems significantly.

Biomass energy in Lech

The popular ski resort of Lech in Austria is situated in a valley. In winter, the pollution caused by the oil heating used in the hotels and other buildings in the village hangs over the valley. A biomass heating plant has just been built to reduce the level of pollution and improve air quality.

The energy to fuel the biomass plant comes from a renewable energy source available in large quantities in Austria – residues from the paper and wood pulp industries. This type of fuel is produced from a natural organic material which is readily available from local sources, such as wood chips, the bark from trees or the sawdust from industry.

The biomass is transported from the conveyor system direct to the biomass oven where it is heated up to 1100°C and distributed as warm water in pipes throughout the village. The outgoing temperature is approximately 100°C and the incoming temperature from the pipes is about 50°C.

It is the first time in Europe that an entire tourist resort has been heated using biomass. The village was offered free connection to the new power plant by the federal and local government as an incentive to dispense with their oil-fuelled heating systems. Currently, 90 per cent of the hotels, households and buildings in the resort have taken up the offer. The biomass heating plant replaces heating that used 3.5 billion litres of oil, which reduces carbon dioxide emissions by about 10 000 tonnes each year. Furthermore, Austria is able to sustain this non-polluting method of heating because of its abundance of trees.

Domestic heating

More than 570 000 homes are currently heated by wood-fired systems, many of which are already equipped with modern combustion technology. Over 70 per cent of biomass is used in low-temperature applications, i.e. the combustion of wood or wood chips in single heaters or central heating boilers in the case of small-scale users, or of various fuels such as bark, sawmill residue, wood chips, or straw in biomass-fired district heating systems.

Further information

Austrian Energy Agency – EVA
Linke Wienzeile 18
A-1060 Vienna
Austria
Tel: +43 1 586 15 24
Fax: +43 1 586 94 88
E-mail: eva@eva.wsr.ac.at
Website: www.eva.wsr.ac.at

Federal Ministry of Science and Transport
Minoritenplatz 5
A-1014 Vienna
Austria
Tel: +43 1 531 20-0
Fax: +43 1 531 20-6480
Website: www.bmwf.gv.at/

Chapter Two
Energy-efficient living

The need to reduce energy use in order to slow the rate of climate change is widely accepted, but the examples in this chapter also illustrate compelling financial reasons to select more efficient technologies and practices.

In Madagascar, fuel-efficient stoves were introduced because of deforestation, but the savings from lower charcoal consumption pay for food and clothing, and manufacture of the stoves has facilitated income generation for local artisans. In China, large businesses and institutions installing compact fluorescent lights have found that labour costs as well as electricity bills are lower, because these lights last longer than ordinary incandescent light bulbs.

In rural Kenyan households, where candles or kerosene lamps are the normal form of lighting, the issue is less one of energy consumption than of convenience. Solar lanterns store energy from the sun in a battery to enable the bulb to be illuminated after dark, so they are cheaper to use and can provide power for a radio, too. In South Africa, pay-as-you-go solar home systems can run lights, radio and television, again by charging a battery during the hours of sunshine.

Also in this chapter are examples of significant reductions in energy consumption in Europe – where the scope for savings is considerably higher than in the South, because of the ever-increasing range of electrical devices in the home and workplace. In Germany, schools have shown how financial incentives encourage effective conservation measures to be adopted, while the Solar House in the United Kingdom has shown how, even in a cool climate, more power and hot water can be generated than are needed by the household.

Like the Solar House, the other innovations described in the chapter show that appropriate building practices can reduce energy use without sacrificing comfort. In Germany, low-energy housing incorporates solar panels and is designed to optimize energy use. In Austria, Passive House standards take this a step further, eliminating the need for heating or air conditioning systems and therefore achieving even greater efficiency.

Reducing household fuel consumption

Madagascar is only 400 kilometres from Mozambique and the African mainland; however, the native wildlife and plants have evolved in a different way from those elsewhere in the world. These unique plants and wildlife are under increasing threat as the villagers scour the forests for fuel to sell to the people in the town, in order to generate income for themselves.

It is often believed that deforestation in Madagascar is caused by slash and burn activity. In the eastern rainforests and tropical forests, the misuse of fire and agro-pastoral systems does contribute to deforestation, but the main pressure comes from the requirement for charcoal, fuelwood and construction wood for supply to urban centres along the coastal areas.

Providing consumers with an alternative, cheaper source of energy is the only practical way to reduce the amount of fuel being used. By reducing the relatively high percentage of the household budget used to purchase the fuel for cooking, families are now able to spend more of their income in other areas, such as nutrition, health, education for children, lodgings and other things that can improve the standard of living for urban dwellers.

The energy project

An energy project has been set up by the World Wide Fund for Nature and a local NGO called the Association to Save the Environment, which focuses on innovative energy-saving techniques. The energy project is producing and marketing more fuel-efficient charcoal-burning and wood stoves for cooking. These stoves have the potential to reduce the amount of natural forest products used in food preparation by between 30 and 50 per cent. While fuel-saving technologies are no substitute for reforestation, they are an important complementary activity.

Improved charcoal-burning stoves

Until recently, in Toliara in south-west Madagascar, 95 per cent of the population used charcoal-burning stoves as their main means of cooking, with many families spending between 10 and 25 per cent of their income on purchasing the charcoal to fuel the stoves. The improved charcoal-burning stove retains heat and reduces the amount of charcoal used by at least 30 per cent. The new design cuts down on deforestation and helps to reduce poverty because less income is needed for the supply of fuel.

The charcoal stoves are produced by local artisans. Although the initial capital cost is quite high, the investment has been recouped by a household after only two months because of the huge savings made on the fuel supply. Users of the new improved charcoal-burning stove find that instead of buying five sacks of charcoal a month, they need only two. The extra money can be used for rice, meat and clothes for the children.

The target was to sell 600 stoves in the first twelve months and 2300 the following year. Within five years, it is hoped that half the families in the region will have been converted to the new improved stoves.

The Kenya Ceramic Jiko

The Madagascan improved charcoal-burning stove has been modified from the Kenya Ceramic Jiko, which is made of a ceramic liner fitted inside a metal case. It burns between 25 and 40 per cent less charcoal than the traditional stoves and it saves about 25 kg of charcoal per month. The Kenya Ceramic Jiko has been widely and easily adopted because it is modelled on the traditional Kenyan metal stove design and requires no change in the methods of cooking or stove use.

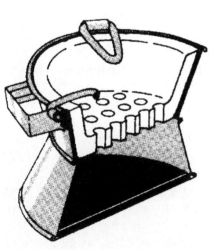

A cross-section of the Kenya Ceramic Jiko

Two basic materials are needed to make the Kenya Ceramic Jiko: clay and metal. These materials should be found close to the production sites and markets to keep transport costs at a minimum.

Properties of clay

The availability of the right clay is the critical factor in determining whether the Kenya Ceramic Jiko can be made. A ceramic liner needs to have physical strength, the ability to resist thermal (heat) shock and insulating capability.

The best type of clay to use in liner making is one with a good physical strength when fired to 900°C; it should remain slightly porous when fired to 1150°C, increasing weight by more than 10 per cent when soaked in water; it should not warp when fired to 1250°C; it should fire to a light salmon colour; it should shrink less than 8 per cent going from a wet plastic state to dry; and it should be plastic.

If the clay does not have these properties, different clay types can be mixed together. Sand, sawdust and other minerals can also be added to form a suitable body. The fewer additional materials that have to be mixed with the clay, however, the lower the production costs and the less supervision required.

There is no substitute for good quality materials and it takes time to develop the right clay mixture. A liner maker must have a well-designed, economical kiln which fires quickly and evenly. Although good quality liners can be made by hand on a potter's wheel, a special moulding machine will significantly improve output, appearance and durability of the finished liner.

Clay deposits

Deposits of clay must be extensive, covering at least 2500 square metres, to a depth of no less than 50 cm. If the deposit is smaller, it cannot be considered a dependable supply. Deposits must be accessible and close to the surface because topsoil that is more than 2 metres deep makes mining expensive and difficult. No more than 50 per cent sand should be present in the clay's natural state.

Properties of metal sheeting

The stove casing is made from mild sheet steel, although galvanized material is used sometimes if suitable supplies of scrap are not available. All the joints in the casing are either riveted or folded, which means that no welding, soldering or brazing are required. A thin steel can be used in the Kenya Ceramic Jiko because the ceramic liner protects the metal from direct contact with the burning coals and any rusting develops slowly. The double cone shape of the casing provides inherent strength, which is reinforced by inserting the ceramic liner.

The minimum recommended thickness of sheet for the casing is 0.5 mm and the maximum is 0.8 mm. A good source of mild sheet steel is scrap bitumen drums. The entire stove can be made from this material, with the exception of the pot-rests, feet and pot-rest holders. The pot-rests should be made from mild steel roundbar, 7–8 mm in diameter, and the feet and pot-rest holders from mild sheet steel at least 0.8 mm thick, but preferably 1 mm. The pot-rest holders can be made from 200-litre oil drum scrap metal, which is usually about 1 mm thick. They are subject to higher loads than other parts of the casing and may tend to unfold if made from thinner material.

Use and care of a Kenya Ceramic Jiko

1. Charcoal should be broken down to an appropriate size before loading into the stove – do not break up large pieces of the charcoal inside the stove. The airgate should be open.
2. Use a mixture of ash and kerosene underneath the stove to light it.
3. Fan the fire after lighting, again with the airgate open.
4. Leave the airgate open if a lot of heat is needed.
5. Close the airgate for slow cooking.
6. To extinguish the fire, shake the ashes and coals out on to the ground, then pour water on to the coals. Do not pour water directly on to the stove to extinguish the fire.

The Kenya Ceramic Jiko has been widely adopted as it reduces charcoal use by 30 per cent
© ITDG/Zul Mukhida

Further information
ITDG
Schumacher Centre for Technology and Development
Bourton Hall
Bourton-on-Dunsmore
Rugby
Warwickshire
CV23 9QZ
United Kingdom
Tel: +44 (0) 1926 634400
Fax: +44 (0) 1926 634401
Email: itdg@itdg.org.uk
Website: www.itdg.org

Energy-efficient lighting

China is responsible for emitting the second largest amount of carbon dioxide and other 'greenhouse' gases in the world. If this output continues to grow at its current rate, China will overtake the USA and become the largest emitter in the twenty-first century. The development of the electric power industry in China has expanded rapidly and since 1978 the output of the total power generation has quadrupled. Shortages of electric power supply and low efficiencies in the use of energy remain serious problems.

Lighting consumes approximately 10 per cent of the total electric power supply in China today. Low-efficiency devices such as incandescent lamps and magnetic ballasts still dominate China's lighting sector, leading to high electricity consumption and environmental pollution. Saving electricity on lighting will alleviate electric power supply shortages that create tremendous lost potential for Chinese industries and will protect the environment. At the same time, energy efficient lighting will reduce peak load and improve the quality of the power supply.

The Beijing Energy Efficiency Centre (BECon) was formally established in December 1993 as a non-governmental, not-for-profit organization to promote energy efficiency and environmental protection in

China. BECon has played an active role in designing, organizing and implementing the China Green Lights Programme.

The China Green Lights Programme

Through its Green Lights Programme, China is attempting to make a significant contribution to reducing global climate change at the same time as working to maintain its economic growth and provide a higher quality of life for its people.

The China Green Lights Programme was begun in October 1996 and aims to popularize high-efficiency lighting products; to save electric energy used for illumination; to promote the manufacture of the new illuminating appliances; to improve the competitive power of enterprises; and to protect the environment. This is to be achieved through encouraging the manufacturing sector to produce high-efficiency, long-life compact fluorescent lights (CFLs), and by communicating to people the financial savings possible from lower electricity bills and the need to reduce 'greenhouse' gases as a society.

Compact fluorescent lights (CFLs)

Compact fluorescent lights (CFLs) are between 50 and 80 per cent more efficient than traditional incandescent light bulbs. An 11 watt CFL can produce an equivalent light to a 40 watt incandescent bulb and it lasts on average eight times as long. Therefore, CFLs can reduce demand by lowering electrical usage and by using a single light bulb instead of eight.

It is essential to maintain high quality in the manufacturing process, because the initial investment in the CFL bulb is about ten times the cost of a normal light bulb. The Chinese Government has created a new national standard for CFLs that at least meets, if not exceeds, international safety and performance standards. The Chinese National Light Testing Centre is preparing to test the products of over 100 of China's manufacturers of CFLs. Those that pass will receive a new certification that should help to convince consumers that they are receiving value for money.

In order to be awarded the certification, manufacturers need to produce CFLs that last for a minimum of 5000 hours. Some very high quality bulbs can last over 10 000 hours. In comparison, traditional incandescent bulbs usually last for around 1000 hours. On average, it will require approximately eight replacements of an ordinary bulb to last as long as a single compact fluorescent bulb.

Energy savings

The energy savings over the useful life of the CFLs are significant compared with incandescent bulbs, and the widespread use of CFLs has the added benefit of reducing the overall energy use – which in turn reduces the emission of climate-changing 'greenhouse' gases.

While each individual who changes bulbs makes a difference, large institutions such as the Chinese Ministry of Railways can have a more immediate and larger beneficial impact. The Ministry is at the forefront of large institutional users in mandating that their facilities use the new and more efficient bulbs. When large institutions, businesses and consumers all join together and start switching to CFLs, the savings can begin to be calculated in terms of the number of power plants not needed.

Businesses have found that they can save money by using the new lighting technology. For example, the Liang Ma Hotel and Office Complex has a much lower electricity bill and its labour costs for maintenance have been reduced because the new bulbs last approximately eight times as long as the traditional light bulbs.

The programme target of 300 million compact fluorescent lamps, small diameter fluorescent lamps, and other high-efficiency illumination products installed in place of conventional lighting, would enable

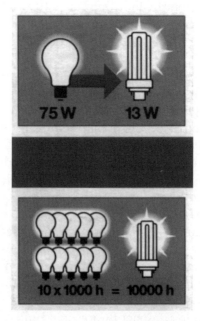

A compact fluorescent light bulb can last more than eight times as long as a conventional bulb

savings of 22 billion kilowatt-hours of electricity at the terminal, leading to reductions in environmental pollution of an estimated 200 000 tonnes and in carbon dioxide emissions of 7.4 million tonnes.

Chinese environmentalists

With the Chinese people representing one-fifth of the population of the world, it is imperative that they take a proactive approach to global environmental conservation in order for it to succeed. While the energy savings and conservation benefits of CFLs are the same worldwide, if the population of China decided to use CFLs instead of the traditional incandescent bulbs, it would mean 1.3 billion people switching bulbs and this would have a hugely positive effect on the environment.

Disposal of CFLs

While the energy savings and the reduction of 'greenhouse' gas emissions are significant, disposal of the bulbs at the end of their life remains a problem. There are traces of rare elements in the CFLs and there is also mercury vapour in some designs, so it is vital that governments, manufacturers and consumers work together to dispose of or recycle the bulbs safely.

Further information

Beijing Energy Efficiency Centre (BECon)
Zhansimen, Shahe
Changping, 102206
Beijing
P.R. China
Tel: +861 6973 2059/6973 5234/6973 3114
Fax: +861 6973 2059
E-mail: becon@public3.bta.net.cn
Website: www.gcinfo.com/becon/becon.html

Solar lanterns

In rural areas of Kenya, many people have no access to electricity and so most families are forced to rely on candles or kerosene lamps to provide basic lighting in their homes. As many as 96 per cent of householders rely on kerosene lamps, and 70 per cent supplement this lighting with battery-powered torches. Kerosene lamps are potentially hazardous, and the costs of fuel and batteries are high, which eats up valuable financial resources. For many people, the provision of light in their homes can compete directly with other household essentials.

Energy from the sun is freely available and many countries in the developing world have it in plentiful supply. As a result there has been a growing interest in the use of photovoltaics (the technology used to convert the sun's energy into electricity) as a renewable and environmentally friendly energy source with which to provide low-cost light and power in rural communities. Unfortunately, the cost of installing even a modest solar home system puts it beyond the means of most rural families in the developing world.

Solar lanterns

The solar lantern, 'Glowstar', designed by Intermediate Technology Consultants (ITC), the consulting subsidiary of ITDG, has been designed as a low-cost alternative to a solar home system and is intended to allow rural families to climb the first step on the 'energy ladder'. The lantern is cheap to maintain and harnesses a free and plentiful source of energy as it is powered by sunshine.

The solar lantern kit consists of a photovoltaic panel and a lantern containing a high-efficiency lamp, a rechargeable battery and a charge control circuit. The lantern uses the

Solar lanterns can be fitted with an output socket to run a small radio
© Phil Webb

latest technology to provide a simple solution to a very basic problem. During daytime, sunlight falling on to the photovoltaic panel generates a small electrical voltage. This is used to charge the lantern's battery so that the lamp can provide light and power during the hours of darkness.

The solar lantern is ideal for any application where there is no local connection to grid electricity, such as rural households and farms, schools and colleges, hospitals, health clinics and other community centres. It also has important applications where there is an inconsistent or unreliable supply of electricity.

Manufacturing options

There is a range of options in terms of manufacturing techniques available to the designer today, and the choice of the most appropriate comes down to considerations of scale of production and cost. As a general rule, individual component part costs come down as production levels increase.

The specified design characteristics include:

- The lantern should give a 360-degree spread of light.
- The bulb enclosure should allow maximum transmission of light with minimum dispersion effects.
- The carry handle should be sturdy and comfortable.
- The lamp should be portable and weigh no more than 2.5 kg.
- The lantern should be stable with a good base.

Potential customers expressed a need for some extra features:

- An indicator to show that the battery is charging.
- A warning light to show that the lamp is about to switch off when the battery is low.
- A power output socket to allow a small radio to be connected to the unit.

Solar power

The photovoltaic solar panel collects light energy from the sun and converts it into electrical current. The energy is stored as charge within the battery of the lantern.

The lamp

The lantern uses a high-efficiency compact fluorescent tube as its light source. These lamps are almost six times more efficient than a standard incandescent light bulb and have an effective life which is eight times longer. In normal use, the lamp can be expected to last for about four years.

Power storage

The rechargeable lead-acid gel-cell is located in the base of the lantern. It has been specially designed to give reliable performance over numerous charge and discharge cycles.

To charge the lantern, the photovoltaic panel is plugged into the clearly marked INPUT socket. Alternatively, if there is a supply of mains electricity nearby then the lantern can be charged using a simple AC adapter. With an appropriate plug and lead the lantern may also be charged through the cigarette lighter socket found in many vehicles.

The OUTPUT socket supplies 9V DC, which is sufficient to power a small radio or cassette player. Both sockets are protected against short circuit and reverse polarity connections.

Suspending the lantern

To maximize the spread of light over an area indoors, the folding hook on the base of the lantern may be used to suspend it, for example in the middle of a room.

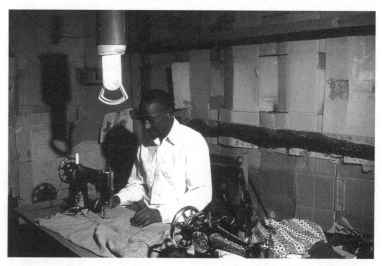

The solar lantern enables rural people to continue working after sunset
© ITDG/Zul Mukhida

Battery technology

A crucial component for any rechargeable device is the battery. The project activities have included research into available battery technologies to identify a battery which:

- has the capacity to store charge sufficient for the required period of lighting (up to four hours)
- is suitably robust to withstand the heavy-duty cycle required for daily charge and discharge
- requires no customer maintenance and is spill- and leak-proof
- has minimum impact on the environment if disposed of at the end of its life cycle
- could be manufactured locally in the medium term in developing countries
- provides a cost-effective solution.

As a result, a Valve-Regulated Lead-Acid (VRLA) battery with a gel electrolyte was selected as the battery technology with which to prototype the lantern. This sealed battery gives a good service life, high energy density and good performance.

Charge control circuit

The charge control circuit housed within the lantern is the 'brain' of the unit. It ensures that the battery is charged and discharged correctly so that it gives a lifetime of maintenance-free service, and it will also give the battery an extra top-up charge if the panel has gone without its full quota of sunlight for a few days. Its on-board microprocessor will even store information, which can be downloaded later, on how the lantern has been used over a period of time. This information is extremely useful and will help the designers to build a picture of how customers use their lanterns, so that they can improve designs in the future.

Further information

Intermediate Technology Consultants
Schumacher Centre for Technology and Development
Bourton Hall
Bourton-on-Dunsmore
Rugby
Warwickshire CV23 9QZ
United Kingdom
Tel: +44 (0) 1926 634400
Fax: +44 (0) 1926 634401
E-mail: glowstar@itdg.org.uk
Website: www.itcltd.com/solar

Note: The ITC Solar Lantern is a UK Registered Design No. 2083358.

Solar home systems

Small, off-grid photovoltaic systems of less than a few kilowatts are ideally suited to the conditions that prevail in the rural areas of South Africa. Eskom, South Africa's national electricity supplier, and Shell Renewables Limited, have embarked upon a joint venture which supplies a unique solar home system to homes in the remote and rural communities of South Africa. This project is the largest commercial solar rural electrification venture ever undertaken. The aim is to bring light to 50 000 rural homes in South Africa.

Solar power is an attractive energy technology because photovoltaic modules produce no pollution, have an expected life of 20 years and require little maintenance. Furthermore, they are now technically proven and commercially available.

Solar photovoltaic cells are manufactured from silicon and assembled into modules that can be installed in a variety of ways to capture the sun's power and meet energy needs.

The system that has been introduced in rural South Africa was designed by Shell Solar and Conlog (Pty) Limited. It has four main components: a solar panel, a charge-controlled battery, a security system, and a metering unit. All of these are prewired and already fitted with special plugs, which is why the system – named the Power

House™ – is also known as 'plug and play': a radio can be plugged in immediately.

Security

There is a special device, the SmartSwitch™, that protects the system from tampering and theft. It prevents the major components, such as the solar panel or battery, from working unless they are connected to their associated control systems.

The system is secure because it cannot be activated without a compatible token or magnetic card. The solar panel and/or battery cannot be used with any other device. The solar panel cannot be connected directly to the battery. The battery is housed in a single enclosure with the security system, charge/credit controller, token reader, termination panel and a prepayment device. This sealed housing unit ensures that only authorized individuals have access to the battery and the controller electronics.

Location

The solar panel has to be positioned at the centre of a wooden pole, outside the home or other buildings, in a place where nothing can block the sun from reaching the panel. For example, it should be situated a good distance from the shadow cast by trees or other developments

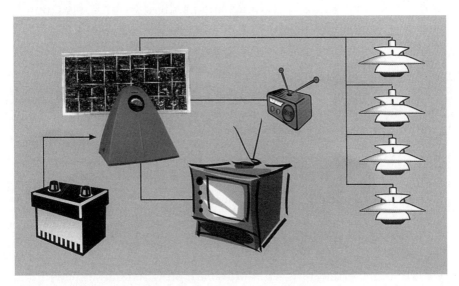

The Power House™ is a totally integrated solar control system
© Shell Renewables

nearby. Consideration also needs to be given to the direction that the panel is going to face, and this depends on the latitude of the country where it is being erected.

Electricity generation

Once the solar panel has been positioned correctly, the photovoltaic cells convert the daylight directly into electricity, which can be used straight away, or can be stored in the battery for use when the sun is not shining. On a cloudy day, electricity may still be generated, although not as efficiently. Heavy cloud and rain will prevent electricity from being generated at all. In these conditions, the household should use the electricity stored in the battery very sparingly. There is an indicator that shows the amount of electricity available in the battery.

The battery can store only a limited amount of energy. Appliances that generate heat, such as irons, stoves, kettles and hotplates, use a large amount of electricity to function, whereas lights, televisions and radios are ideally suited for solar energy because they require only a small supply of electricity to operate.

Costs and maintenance

The system is compact and easy to install. It is activated by a magnetic card, which is paid for in advance and credited with 30 days' power supply. This method guarantees revenue collection. If a payment is not made, then the system becomes disabled. The monthly payment also covers maintenance, repair and replacement of the equipment and free exchange of light bulbs.

Once the initial installation cost has been met, customers have found that the magnetic cards are no more expensive than candles and paraffin but provide greater energy efficiency and are more sustainable. The prepayment system gives people an opportunity to install solar power without having to make a large capital investment in equipment that they might not be able to afford.

The Power House™ has been designed to be a customer-friendly system. There is a simple electronic display that indicates the number of days of usage remaining before the load is disconnected, as well as the charge remaining in the battery.

Further information

Shell Solar Energy B.V.
PO Box 849
5700 AV
Helmond
The Netherlands
Tel: +31 492 508608
Fax: + 31 492 508600
E-mail: info@shellsolar.nl
Website: www.shell-renewables.com

Shell Solar Southern Africa (Pty) Ltd
PO Box 2703
Pavillion
3611
South Africa
Tel: +27 31 265 1260
Fax: + 27 31 265 1263
E-mail: meterexp@iafrica.com

Conlog (Pty) Ltd
PO Box 2332
Durban
4000
Kwazulu Natal
South Africa
Tel: +27 31 268 1111
Fax: +27 31 209 5073
E-mail: corporate@conlog.co.za
Website: www.conlog.co.za

Note: Power House™ is a registered trademark of Shell International.
SmartSwitch™ is a registered trademark of Conlog (Pty) Ltd.

Schools energy-saving project

In October 1994, it was decided that schools in Hamburg were using too much energy. In an attempt to conserve some of the energy that was being wasted, the Fifty-Fifty Project was started in a number of schools. The key element of the project is a system of financial incentives that enables the schools to share the saving in energy and water costs they have achieved. Fifty per cent of the money saved in energy conservation is returned to the school, where it can be reinvested in new energy-saving devices, equipment, materials and extra-curricular activities. For instance, Blankenese School bought solar panels with the money they saved on energy consumption and installed them themselves.

Special equipment was installed to monitor the progress of the project in the schools. Each school was responsible for recording its progress and the results were monitored by the Environmental Agency. Each school had to develop its own methods of carrying out the work within the constraints of its facilities.

It was intended that the trial project would last for three years in all types of school, with 24 schools taking part initially. By July 1995, 40 schools were involved and by the autumn of 1996, 60 more schools had joined the project. Since January 1997, the project has become a permanent exercise and the savings have continued to increase.

Implementation of the Fifty-Fifty Project

A team was set up in each school – teachers, cleaners and a member of the staff management team. Their job was to inform the school about the project and their plans to implement it. The members of the team reported back to the Environmental Agency and ideas were brought together from all the schools, including the problems they faced.

Some schools set up teams of pupils – the 'Energy Team', the 'Lighting Team', the 'Electricity Team', etc. to carry out patrols to monitor energy waste and water consumption. Other schools organized teams out of school hours to carry out the work. The exercise encourages the students to think about energy waste and eventually they will save water and energy automatically.

In most cases, the teachers initiated the ideas and monitored the procedure. Many of the ideas were integrated into lessons – some subjects lent themselves more readily to this work, but in art, for example, posters were made to advertise the project.

The first year was the hardest. Everyone needed to be organized, determined and committed to the project. Results were not always immediately obvious but once the project was up and running properly results soon began to show. The credit goes largely to the commitment of the teaching staff who made this project the success it has been, though each school needs the full support of its caretaker.

Saving electricity

One school removed all its hot water boilers and saved money in this way – remarkably, no one complained! In the staff room one of the two fridges was switched off as it was hardly used, and the students were trained to switch off lights at break or when lessons were cancelled. Coffee was kept warm in flasks instead of keeping the coffee machine running. All the electrical appliances, such as fax machines and photocopiers, were switched off at the end of the school day.

Saving heat

Another school decided that their rooms were generally too hot, so they reduced the water temperature for the radiators from 70 to 50°C and found their working conditions to be more pleasant. Thermostats were checked regularly. Doors and windows were closed at the end of the school day to conserve the heat.

The heating system should be checked on a regular basis with the room temperature kept at 20°C, ensuring that it does not fall below 10°C. When the heating level rises and becomes too high, the energy patrol asks the caretaker to turn down the centrally regulated system.

An empty classroom is always much colder than a full one, so as the temperature inside the room increases the heating should be turned down, rather than opening a window and letting out all the hot air.

Saving water

In the first year, the 24 schools taking part in the project saved 7941 cubic metres of drinking water, which is the equivalent of the annual consumption of 80 households. However, half the schools involved found this part of the project the most difficult. Most of the water was used for toilets and for showers after sport so there was little opportunity for reducing the consumption. Instead, the students were encouraged to report dripping taps or leaks and the caretaker was encouraged to check the toilet flushes regularly.

Swimming pools in schools waste a lot of water, so the level of the water in the pools was lowered in an attempt to reduce the amount lost over the side from splashing.

Savings have continued to increase each year since the beginning of the project.

The future of the Fifty-Fifty Project

The initial costs incurred for this project by the City Council of Hamburg were considerable, but were balanced by the great savings in energy and water made by the schools. The success of the scheme depended very much on the exchange of knowledge and experiences in implementing the project and the working together of the various groups. Many more schools have enquired about the project throughout Germany and there are now 420 taking part, with a quarterly periodical reporting on successes.

The Fifty-Fifty Project is now being established at schools in Japan, Greece and Spain. The savings can be huge and can amount to an average of US$3000 per year being put back into each school. The project has now been extended to include waste disposal.

By saving energy, the schools are reducing carbon dioxide emissions and so the environment is protected. The results in the trial period show that, by the year 2005, the Fifty-Fifty schools will have reduced the emissions of carbon dioxide by 25 per cent. It has been found that students are more interested in this aspect than in saving money, and so the approach has been adjusted to reflect their enthusiasm for slowing climate change.

Further information

Fifty-Fifty Projekt
Wolfgang Thiel
c/o Umweltbehoerde i6L5
Billstrabe 84
20539 Hamburg
Germany
Tel: +49 40 42845 2225
Fax: +49 40 7880 2099
E-mail: Wolfgang.Thiel@ub.hamburg.de
Website: www.hamburger-bildungsserver.de/klima/fifty-fifty

The solar house

Until recently, the majority of buildings in the United Kingdom were planned without considering the energy demand. The depletion of fossil fuel reserves and the threat posed by this to our environment has now led to alternative sources of energy in buildings being considered.

The integration of photovoltaic modules into the structure of a building can have a number of benefits besides electricity production, including the use of the solar elements as a roof covering or shading structure. Currently, there is little demand in the United Kingdom, but cost reductions, climate change, cultural changes and developments in technology are beginning to increase the potential of photovoltaics as a viable contributor to power demand.

The Oxford Solar House is the first low-energy house in the United Kingdom with a fully integrated photovoltaic roof. The house was designed to function as an ordinary standard family home, which requires only a minimum amount of energy for heating, cooling and lighting. In order to optimize the value of the electricity generated by the photovoltaic system, the energy demand in the house was reduced by using all available energy-saving technologies but without impairing the comfort of the occupants.

The house was built to evaluate the potential for photovoltaics to contribute cost-effectively to domestic and industrial energy supply, and to demonstrate the potential of solar energy to replace as much as possible the environmentally damaging electricity and gas supplies in a dwelling, which result in carbon dioxide emissions and so contribute to climate change.

Location

The construction of the house took about 18 months and it was completed in March 1995. The house is orientated roughly east–west with a south-facing rear elevation that provides good solar access. The house receives approximately four peak sun hours in summer, but only 0.6 peak sun hours in winter (Dichler, 1994). During the summer months, energy surpluses are predicted to be around 12 kWh per day, which is greater than the house energy deficit in winter. The house, therefore, has a positive energy balance. Power has to be drawn from the utility during night-time and winter days.

Design of the solar house

The house is laid out with rooms arranged round a central core, incorporating a service duct, stairs to the first floor and a hallway to the entry porch. Bathrooms are positioned over the kitchen to reduce the length of pipe needed and hence material used. The front and back doors are protected by a porch to the north and a two-storey double-glazed conservatory to the south, with a balcony on the first floor between two bedrooms.

Warm air is taken in from the conservatory air space through ground- and ceiling-level vent windows and French doors. It circulates through the house by convection to the kitchen and upstairs bathrooms, where it is expelled through windows. Some (one-third) of the sunspace is roofed to prevent overheating in the summer.

Gas-fired appliances are used for cooking all year round and in the winter months they are used to preheat water for the washing machine and dishwasher, and for heating three radiators in the north-facing rooms for two hours a day. The use of gas appliances removes a potentially large electricity burden that would normally be connected to the utility supply. At ground level, a wood-burning Kakkleoven is the main source of heating. The walls, windows, floor and roof are well insulated to ensure low heat loss and there is triple glazing throughout the house except in the sunspaces. Materials were chosen carefully for transport energy, durability and heat storage.

South-facing rear elevation of the Oxford Solar House
© Susan Roaf

Heating and hot water systems

The north-facing rooms have a central heating system installed, while the larger south-facing rooms rely on solar gain and conservatory pre-heating. The efficient wood-burning Kakkleoven on the ground floor supplements solar gain during winter. The greatest fluctuation in temperature on the ground floor occurs as a result of doors being opened.

The house has a 5m² flat plate solar collector which is mounted on the roof co-planar with the photovoltaic system. The solar hot water collectors are used to supplement the energy demand for domestic hot water and to supply about 77 per cent of the household requirements, including the washing machine and dishwasher (Viljoen, 1995).

The heating demand is reduced by maximizing passive solar gain, by providing thermal mass to even out temperature swings, and through good insulation. The gas energy demand for hot water is minimized by installing a solar hot water system. The grid electricity demand is reduced by installing energy efficient appliances, and by using a photovoltaic array mounted on the roof.

Ventilation

The house has no mechanical ventilation system but there is no condensation because the air and wall temperatures are typically the same. The house has a wide range of tilt-and-turn windows and vents to the sunspace which prevent it from becoming stuffy.

Photovoltaic system

The photovoltaic system is connected to the electricity supply and is regarded as the main power supply. It was designed to export surpluses to the National Grid, importing power only when it is essential and unavoidable, such as night-times and overcast conditions. Low-consumption appliances and careful timing in use are essential to spread the electric loads.

The photovoltaic modules (4 kW) are mounted between the skylights of the roof. Edge frames have been used and the modules pre-carried on an aluminium substructure mounted to the roof. The frames are designed to fasten, by simple means, to as standard a roof structure as possible.

Economic performance

One of the most attractive aspects of roof-integrated photovoltaic systems is that the modules can be used in place of a conventional roof. Building integration of photovoltaics leads to reduced costs for infrastructure, installation and groundwork; savings in materials; lower installation and planning costs; and integrated maintenance and operation.

Total cost of the installation of the photovoltaic system	£21,150
Average annual exported electricity	1692 kWh
Average annual imported electricity	1524 kWh
Average annual electricity consumption (from previous estimations)	2964 kWh

The Oxford Solar House produces only 140 kg of CO_2 per annum, compared with 6000 kg of CO_2 emissions from the other houses in the same street. Not only is there enough power generated to sell back to the electricity company, but there is a sufficient amount to power an electric car. The car takes three hours to charge to travel 30 kilometres.

Further information

Susan Roaf
School of Architecture
Oxford Brookes University
Headington
Oxford OX3 0BP
United Kingdom
Tel: +44 (0) 1865 483 200
Fax: +44 (0) 1865 483 298
E-mail: sroaf@brookes.ac.uk

Urban energy policy

Freiburg is located at the heart of the warmest and sunniest region of Germany, and its city council has implemented a policy to make greater use of solar power. The Freiburg energy supply concept focuses on:

- energy saving projects
- use of renewable energy sources
- application of innovative energy technologies
- extension of on-site and remote district heating systems.

In June 1992, the Freiburg city council adopted a resolution to the effect that it would only permit construction of 'low-energy' buildings on municipal land. All new housing must comply with the low energy guidelines. Low-energy housing uses solar power passively as well as actively. In addition to solar panels and collectors on the roof, providing electricity and hot water, many passive features use the sun's energy to regulate the temperature of the rooms.

Water heating technologies are usually referred to as active solar technologies, whereas other technologies, such as space heating or cooling, which passively absorb the energy of the sun and have no moving components, are referred to as passive solar technologies.

The whole city is involved in Freiburg's solar policy. Many private companies and public facilities make their roofs available for solar modules. The people of Freiburg buy shares in the panels and are reimbursed when the power is sold to the city electricity scheme.

Passive thermal use of solar energy

There are many applications for the direct use of solar thermal energy; for example, space heating and cooling, water heating, crop drying and solar cooking. It is a technology that is well understood and widely used in many countries throughout the world. Solar thermal technologies have been in existence in one form or another for centuries and have a well-established manufacturing base in most sun-rich developed countries.

The most common use for solar thermal technology is for domestic water heating. Hundreds of thousands of these domestic hot water systems are in use throughout the world, especially in areas such as the Mediterranean and Australia, where there is high solar insolation (the total energy per unit area received from the sun). As world oil prices fluctuate, it is a technology that is rapidly gaining acceptance as an energy-saving measure in both domestic and commercial water heating applications, although currently domestic water heaters are usually found only among wealthier sections of the community in developing countries.

Passive use can be made of solar energy through structural measures, such as windows and conservatories. It is possible for a building to be partially heated in this way. The solar energy is stored in solid parts of the building, such as walls, ceilings and floors.

Windows can make use of solar energy effectively and inexpensively. The solar energy penetrates the building virtually unimpeded via the window. Solar energy gained in this way is retained inside and stored in solid parts of the building. A window thus acts as a solar collector. To ensure that the heat is not lost via the windows again, however, these require high-quality glazing offering good thermal protection, such as double or triple glazing with heat-absorbing glass. Wood is the best frame material. A good solar house has a small number of big windows rather than a large number of small ones. Walls facing away from the sun (north, north-west or north-east in the northern hemisphere) should have relatively small window surfaces (less than 15 per cent). Sun-facing walls should incorporate a large number of windows – a window area of 25 to 50 per cent has proved successful here.

Conservatories on the south- (or equator-) facing side of a building also permit passive use to be made of solar energy. The glass annexe is heated up by the sun and thus forms an intermediate thermal 'buffer'. This reduces the heat loss via the adjacent house wall, and the heated air in the conservatory can be used to heat and air the main building. Conservatories should be fully thermally insulated from the main building and equipped with high-quality glazing (using at least insulating glass and preferably heat-absorbing glass).

Low-energy dwellings incorporate highly effective heat insulation for all exterior components (including efficient windows) which reduces the heat loss considerably.

Photovoltaic use of solar energy

Solar photovoltaics (PV) is a technology that converts sunlight directly into electricity. PV systems have an important function in areas remote from an electricity grid, where they provide power for water pumping, lighting, vaccine refrigeration, electrified livestock fencing, telecommunications and many other applications. With the global demand for reductions in carbon dioxide emissions, PV technology is also gaining popularity as a mainstream form of electricity generation. Some tens of thousands of systems are currently in use, yet this number is insignificant compared to the vast potential that exists for PV as an energy source.

Photovoltaic modules provide an independent, reliable electrical power source at the point of use, making it particularly suited to remote locations. PV systems are technically viable and, with the recent reduction in production costs and increase in conversion efficiencies, can be economically feasible for many applications. The solar cells on the market today can convert up to 17 per cent of the solar energy that shines on them into electricity.

Applications remote from the grid

Consumers located a long way from the grid can be supplied with electricity by a photovoltaic system. Lead-acid batteries are used to store the electricity. The solar installation should always be combined with energy-saving appliances; for example, energy-saving light bulbs. In stand-alone operations, PV systems are already able to compete with other electricity sources (mains connection, diesel generators) in economic terms.

Grid-connected PV systems

Photovoltaic systems can also be connected to the public electricity grid via an appropriate inverter. No means of storing the electricity is then required. On sunny days, the solar generator will supply the electric appliances in the house, and surplus energy will be fed into the grid. During the night, and on days with unfavourable weather conditions, electricity will be taken from the grid. When planning a PV system, it is important to avoid having modules that are partially in the shade. Where possible, the solar modules should be positioned so that any shade occurs only during periods of weak sunshine (mornings, evenings and winter).

Active thermal use of solar energy

Solar collectors are used to convert solar radiation into heat by means of the black solar absorber inside the collector. A heat transfer medium (generally water, although air is used as well) is conducted through the absorber, heating up in the process, and transports the energy to the point of consumption or storage.

A good solar installation can provide about two-thirds of the energy required, over a year, to supply a household with hot water.

Collector installations

Roof-mounted: The collector is installed above the tiles and the roof remains intact.
Roof-integrated: The collector replaces the tiles, permitting savings on tiles, and is incorporated and sealed in the roof skin (with the aid of prefabricated installation systems).
Free-standing: The collector is installed on a rack on the ground or on a flat roof at a specific angle of inclination.

The photovoltaic modules used in Freiburg are manufactured by the local company, Solar-Fabrik. They are highly efficient, and easy and quick to mount. The high product quality of Solar-Fabrik modules has the following features:

- 15 cm single-crystal silicon cells with anti-reflection coating which gives top conversion efficiencies.
- The modules are reliably protected against thermal expansion damage and moisture penetration.
- The highly transparent and specially tempered glass front ensures excellent light transmitting capacity and robustness.
- The modules connect readily with all inverters commonly found on the market.
- 10-year output warranty.
- Mounting hardware and all other system components are available.

Further information

The International Solar Energy Society (ISES)
International Headquarters
Villa Tannheim
Wiesentalstr. 50
79115 Freiburg
Germany
Tel: +49 761 45906 0
Fax: +49 761 45906 99
E-mail: hq@ises.org
Website: www.ises.org

Solar-Fabrik GmbH
Munzinger Str. 10
D-79111 Freiburg
Germany
Tel: +49 761 4000 0
Fax: +49 761 4000 199
E-mail: info@solar-fabrik.de
Website: www.solar-fabrik.de

Freiburg Municipal Utilities (FEW) –
Energy advisory service
Leopoldring 7
D-79098 Freiburg
Germany
Tel: +49 761 279 2409

Passive House standards

Even in colder regions of the world, it is possible to reduce the energy demand for heating. Cepheus (Cost Efficient Passive Houses as EUropean Standards) – a project involving the construction and scientific evaluation of 250 housing units built to Passive House standards in five European countries – began in 1997 with the following goals:

- to demonstrate the technical feasibility (in terms of achieving the targeted energy performance indexes) and cost-effectiveness of an array of different buildings and designs implemented by architects and developers in a variety of European countries
- to study investor–purchaser acceptance and user behaviour under real-world conditions for a representative range of buildings
- to test the implementation of the Passive House quality standard throughout Europe with regard to cost-efficient planning and construction
- to provide opportunities for both the lay and the expert public to experience the Passive House standard hands-on at several sites in Europe
- to give development impulses for the design of energy- and cost-efficient buildings and for the further development and accelerated market introduction of individual, innovative technologies compliant with Passive House standards
- to create the preconditions for broad market introduction of cost-efficient Passive Houses
- to illustrate the potential of the Passive House standard to meet the energy requirements of new housing in a manner that is both cost-efficient and, in sum over the whole year, produces zero greenhouse gas emissions.

A Passive House is a building with an extremely low heating energy demand. The term 'Passive House' refers to a construction standard, which can be met using a variety of technologies, designs and materials. It is a refinement of the low-energy house (LEH) standard. The Passive House standard offers high comfort, minimal energy consumption and negligible heating costs.

ENERGY-EFFICIENT LIVING   CHAPTER TWO   **73**

No heating or air conditioning system

Passive Houses are buildings in which a comfortable indoor climate can be achieved in winter and in summer without needing a conventional heating or air conditioning system. A separate heating system is not necessary because heating energy can be distributed in the building by means of the existing ventilation system.

The heat inputs are delivered externally by solar radiation through the windows and internally by the heat emissions of appliances and occupants. These inputs are essentially sufficient to keep the building at a comfortable indoor temperature throughout the heating period. The minimal heat requirement can be supplied by heating the supply air in the ventilation system.

Efficient technologies are also used to minimize the other sources of energy consumption in the building, most notably electricity for household appliances. The total final energy consumption for space heating, domestic hot water and household appliances of a Passive House is lower by at least a factor of four than the specific consumption levels of new buildings designed to the standards presently applicable across Europe.

Passive solar gain

South-facing Passive Houses are also solar houses. The passive gain of incoming solar energy through windows which are dimensioned to provide sufficient daylight, covers about 40 per cent of the minimized heat losses of the house. The houses have high insulation through the use of improved windows with high-quality glass and frames that are airtight. The triple glazing and superinsulated frames let in more solar heat than they lose. Argon is the filling material between the layers of glass and there is a special wooden frame construction which has some ventilation between the wooden outside and the wooden inside of the frame. The benefit can be enhanced if the main glazing areas are oriented to the south and are not shaded.

Superinsulation

Passive Houses have an exceptionally good thermal envelope, preventing thermal bridging and air leakage. To be able to dispense with an active heating system while maintaining high levels of occupant comfort, it is essential to observe certain minimum requirements concerning insulation quality.

Combining efficient heat recovery with supplementary supply air heating

Passive Houses have a continuous supply of fresh air, optimized to ensure occupant comfort. The flow is regulated to deliver precisely the quantity required for excellent indoor air quality. A high-performance heat exchanger is used to transfer the heat contained in the vented indoor air to the incoming fresh air. The two air flows are not mixed. On particularly cold days, the supply air can receive supplementary heating when required. Additional fresh air preheating in a subsoil heat exchanger is possible, which further reduces the need for supplementary air heating.

Electric efficiency means efficient appliances

By fitting the Passive Houses with efficient household appliances, hot water connections for washing machines and dishwashers, airing cabinets and compact fluorescent lamps, electricity consumption is also dramatically reduced. Energy use can be cut by half compared with the average housing stock, without any loss of comfort or convenience. All building services are designed to operate with maximum efficiency. High-efficiency appliances are often no more expensive than average ones but, as a rule, they pay for themselves through electricity savings.

Cost-optimized solar thermal systems can meet between 40 and 60 per cent of the entire low-temperature heat demand of a Passive House. The remaining energy consumption (for space heating, domestic hot water and household electricity) can be offset completely by the use of renewable sources.

Principles of Passive Housing

What makes the approach so cost-efficient is that it uses those components of a building that are necessary in every case; for example, the windows and the automatic ventilation system. Improving the efficiency of these components, to the point at which a separate heat delivery system can be dispensed with, yields the savings that largely finance the extra costs of improvement.

Passive houses can be built cost-efficiently. The total costs (including planning and building services, plus running costs over a period of 30 years) in a Passive House are no higher than for an average new building.

Benefits of passive energy-saving technologies

No further elements are required in addition to a conventional building. It is only necessary to construct the components that are used in any case, such as floors, outer walls, windows, roofs and ventilation, to better quality standards than usual. Over the medium term, such a quality improvement need not cause higher investment costs than in a standard house. Particularly through the prefabrication of high quality exterior building elements, such components can be produced very cost-effectively.

Further information

The Energy Institute
Stradstrasse 33/CCD
A-6850
Dornbirn
Voralberg
Austria
Website: www.cepheus.de

Chapter Three
Recycling a valuable resource

Waste reuse and recycling is part of the culture of most developing countries. Waste collectors go out in search of reusable articles, and high levels of skill and ingenuity are brought to the task of reprocessing the waste. Recycling artisans have integrated themselves into the traditional marketplace and have created a viable livelihood in the sector. By contrast, industrialized countries have set targets and other incentives in an attempt to reduce the amount of waste that is simply discarded, using up all the landfill sites and creating potentially explosive methane.

One businessman in Bangladesh has adapted rickshaws and employs sweepers and refuse collectors to clear the rubbish from local streets that are too narrow for the corporation trucks to pass through, in a bid to improve the local environment.

In Pakistan, where there are waste dealers or 'kabaris' who operate from shops or warehouses, the waste passes through many hands as it is collected and sorted and sold on before being recycled. In Denmark householders have been encouraged to sort their rubbish before collection so that it can be more easily recycled.

Waste tyres and plastic are increasingly becoming a nuisance as they do not decompose readily, but a company in the United Kingdom has begun recycling tyre inner tubes to make bags and household accessories, while in China plastic is being made into fuel. A Norwegian company has produced automated drinks container recycling machines to encourage people to dispose of the empty container in an environmentally friendly way.

In Uruguay, empty drinks cans are a particular problem on the beaches, and in this case manual collection has been encouraged by payment for the crushed cans, as recycling them saves the production company energy as well as raw materials.

Natural wastes such as algae and water hyacinth are just as inconvenient when they multiply rapidly as they do in Italy and Uganda respectively, but both have been used for papermaking, and the water hyacinth is used in crafts and furniture making. In India, rags are used for papermaking to reduce deforestation and provide employment.

Finally, shipping containers have been adapted and joined to make community buildings, schools and shops in South Africa, again using skill and ingenuity to adapt unwanted items to provide valuable resources.

Household sorting of domestic waste

The Danish government aims to downsize the rubbish tips in Denmark by ensuring that all the municipalities recycle at least half of their refuse. The object is to send as little as possible to the tip, using it only as the last resort. The reason for reducing the refuse dumped on the tip to an absolute minimum is that nearly all of the waste can be used as a resource.

Fredericia is a town at the very heart of Denmark and is well known for its heavy industry. At the beginning of the 1980s, Fredericia was subject to considerable criticism on environmental grounds. The people of Fredericia have now implemented a system which recycles most of their refuse and contributes to a better environment. The sorting of refuse is the central point of their recycling system and everybody, whether they live in a house or in a block of flats, takes part by sorting their rubbish into different categories. Pamplets are available that provide instructions for people on how to sort out their waste for recycling.

The refuse system for blocks of flats

Environmental pavilions have been placed near all the blocks of flats in Fredericia and the residents are responsible for taking their refuse there and sorting it out themselves. Different bins are provided for the sorting of green kitchen refuse; glass; paper/cardboard; plastic bottles; and expanded polystyrene/metal.

The refuse system for houses

Residents in houses can dispose of their refuse in a number of ways. Green kitchen waste can be composted at home in rodent-proof compost bins provided by the council. The addition of special compost worms – turbo worms – which are also provided by the council, speeds up the process of decomposition. For the people who cannot make their own compost, a central composting site has been established for treating green kitchen refuse and the muncipality collects it every fortnight from large ventilated bins. The compost ends up in gardens and fields to benefit the plants and the environment.

Other household refuse, such as meat and fish remains, milk and juice cartons, nappies, dust from the vacuum cleaner, dirty tins and cans, and leftover food, is wrapped in disposable containers to avoid bad smells and then is collected for incineration every fortnight. Home owners must choose a bin that meets the demands of their household. The refuse collected is incinerated and used to provide heating for many homes.

Recycling centre

The range of wastes that can be recycled is vast: newspapers; cardboard; mixed paper; plastics – for example empty containers from detergents and shampoos; glass – for example bottles, jam jars and chipped drinking glasses; metals – for example tins, cans and bottle tops; green kitchen refuse such as cut flowers, house plants, herbs, fruit and vegetable remains, nutshells and eggshells, bread and cake remains, cereals, coffee grounds and tea leaves; inflammable refuse; construction waste; clean soil; recyclable items; garden refuse; and hazardous waste, such as oil, chemicals and batteries.

Recyclable materials are sorted and treated at the recycling centre. Lids have to be removed from plastic containers and plastics, metals and glass should be rinsed and cleaned prior to disposal. Glass is sorted into two categories – glass that can be remelted and non-disposable bottles. Paper and cardboard are pressed into bales and sent to factories where they are used for making recycled paper and cardboard. Expanded polystyrene is turned into filler material or insulation, or is used for making new expanded polystyrene packaging material. Metal is sent to factories for remelting. Plastic bottles are used for making new plastic products.

Hazardous waste

The municipality has made agreements with pharmacies and paint dealers for the return of medicine and paint remains. Medicine remains can be returned to the pharmacies. Syringes and needles must be put into approved needle boxes which are supplied by the pharmacies. Paint remains can be returned to the paint dealers in tight, closed containers. The paint dealers do not accept mercury, acids, alkalis, vegetable poisons and pesticides, and these materials have to be disposed of at the recycling sites.

Batteries can be disposed of in all places where batteries are for sale. This also applies to rechargeable batteries containing harmful heavy metals such as cadmium and nickel. Car batteries can be returned to the car dealers or disposed of at the recycling sites.

Older and disabled people who need help to return their hazardous waste to the dealers or dispose of their garden refuse can order collection from the council through a telephone hotline number.

Recycling shops and workshops

In Fredericia, most types of household goods are used over and over again. Any items that can be resold are sent to council-run workshops where a team of people repairs and renovates anything that is retrievable before putting them up for sale in the special recycling centre shop. Old clothes are sorted, washed, repaired and sold in the recycling shops. The system could work anywhere because it is low tech and operates with the cooperation of local people.

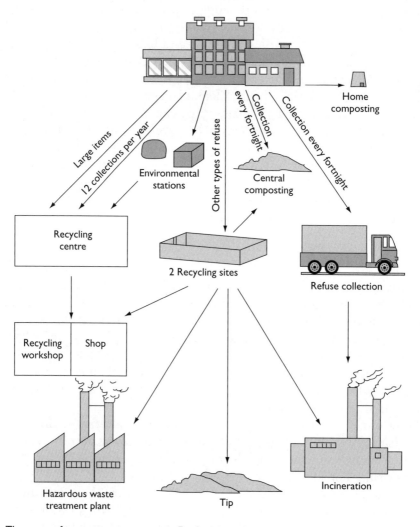

The new refuse treatment system in Fredericia

Recycling success

The municipality has achieved a high level of recycling. As a result, waste combustion has been greatly reduced, and far less waste now ends its days at waste disposal sites.

Cleaner industry and cleaner water

The amount of sulphur dioxide in the air has been greatly reduced because most urban districts are now supplied with district heating which uses surplus energy from the factories. As a result, emissions to the waters of the Little Belt inlet and pollution from factory chimneys have fallen dramatically.

The vast majority of companies in Fredericia have installed flue gas cleaning systems and large companies have started to use natural gas instead of oil. Unlike oil and coal, natural gas does not emit sulphur dioxide on combustion and 99 per cent of the emissions consist of water vapour only.

All waste water from the town is treated at the Fredericia Central Treatment Plant. Effective treatment has led to a significant drop in the amount of pollution emitted to the Little Belt, and the quality of sea water has improved as a result. Household waste in the municipality is now graded so effectively that Fredericia has one of the best waste treatment systems in the whole of Denmark.

Further information

Municipality of Fredericia
Radhuset
Gothersgade 20
DK-7000
Fredericia
Denmark
Tel: +45 79 21 21 21
Fax: +45 79 21 21 20

Informal recycling of waste materials

Increases in population and migration into cities have created serious environmental problems, including inadequate solid and liquid waste management, lack of safe water and minimal pollution control. Many Southern cities are characterized by overcrowded housing, contaminated water supplies and lack of proper sewage disposal, drainage or waste collection, all of which contribute to an unhealthy urban environment. Communities living near dump sites also suffer the nuisance of smoke and smells, and such sites – as well as uncollected waste in general – attract rodents and flies which provide a transmission route for disease.

Cities in developing countries have to deal with increasing quantities of waste – items that are generated and discarded as rubbish by households, commercial and industrial institutions, and hospital waste. Developing countries produce on average 300–600 g of municipal waste per person, per day. For many cities in the South, uncollected solid waste has become a major health hazard, yet municipal waste management services may collect only as little as 25 per cent of the total refuse produced.

Solid waste management in Pakistan

Solid waste management problems in Pakistan increased after the Second World War due to the use of disposable items such as plastic bags which cause drainage problems. The Solid Waste Management (SWM) Department of Karachi Metropolitan Corporation estimates that only half of the city's daily generation of 7000 tonnes of rubbish is collected from the streets by the municipal service, while the rest remains at collection points and on dump sites. As the urban environment in Pakistan continues to deteriorate, there is growing recognition of the need for a sanitation policy and sound operational strategies for dealing with the problem.

The formal sector comprises the government agencies that provide SWM services, such as the municipal service; informal sector activities are those that are not regulated and controlled by government agencies, either in the form of the recycling enterprises of itinerant waste buyers and dealers or through self-employed (private) and municipal sweepers collecting solid waste against an agreed payment from households.

Informal recycling of domestic waste can be divided into two broad categories:

- waste picking in streets, communal bins, transfer points and disposal sites
- waste separation at the household stage and selling on to itinerant waste buyers.

In both cases, the waste passes through a number of dealers involved in trading and recycling before ultimately reaching the recycling industry. All the activities provide jobs or additional income to a significant number of people.

The waste management system starts from the households and ends with the disposal or reuse of the materials. Municipal waste collection begins with sweeping and kerbside collection. This waste is then removed using handcarts or tricycles to large on-road collection points where it is transported by vehicles to the disposal sites.

Major quantities of resaleable waste components are separated either at source or when the waste is transported for disposal. The material is then sold at the household level, with payment made on the basis of the weight of each individual material. It is traded and recycled through an established network of waste pickers, dealers, wholesalers and recyclers. Alternatively, primary waste is removed from the households by sweepers and disposed at transfer points (where the transfer of waste from small collection vehicles is made to larger transport equipment).

The informal sector

There are independent operators dealing in waste collection, purchase, separation, restoration, resale and recycling, with the scale of operations ranging from itinerant manual workers to large recycling factories.

Kabaris are large-scale waste dealers who operate from shops and warehouses. There are approximately 1000 in Karachi and most specialize in just one type of waste, which they buy at auctions or from middle dealers and then recycle or sell to recycling plants.

The Safai Kamai Bank operates every Tuesday from a bazaar in Karachi and uses the slogan 'garbage is gold'. People can bring their dry garbage for sale on a per kilogram basis – the price paid depends on the item. Items purchased include newspapers, other paper waste, plastic bags, metal, glass and plastic bottles.

Waste busters collect rubbish from households and make a small charge each month, which covers the delivery of about 30 rubbish bags. The refuse is taken to the transfer station where it is sorted out and loaded on to trucks for recycling.

A local NGO, Pakistan Environment Welfare and Recycling Program (PEWARP), has established a small production unit manufacturing three organic products from waste purchased from itinerant buyers at Karachi's huge vegetable market. The vegetable waste is crushed and the liquid extract collected, which results in liquid concentrate sold as a pesticide, dilute liquid sold as fertilizer and solid residue.

Shehri, a Karachi-based NGO also known as 'Citizens for a Better Environment', is primarily concerned with the protection and conservation of the natural and built environment. It has produced recommendations for improved bin designs and promotes awareness on solid waste management.

Recycling waste materials

The separation practices are well established and, as a result, quantities of certain waste components, such as bottles, newspapers, plastic, food waste and aluminium cans, are considerably reduced in the waste stream. Once resaleable waste components have been separated from waste they are considered to be raw materials.

Uses for common waste materials

Waste material	Common reuse and recycling
Broken glass	Glass bottles
Bottles	Washed and used again
Bread	Livestock feed
Newspapers	Various types of packing
Ferrous metal	Recycled in re-rolling mills
Paper	Cardboard, etc.
Aluminium	Re-melt in moulds for various industries
Plastics	Uses/recycling depends upon type: toys, shoe soles, shopping bags, sandals etc.
Plastic bags	Buckets and other household containers
Magazines, books	Sold again at reduced prices
Old furniture	Sold again at reduced prices

Further information

Water and Sanitation Program – South Asia
World Bank
PO Box 1025
Shahrah-e-Jamhuriat
Ramna (G-5/1)
Islamabad
Pakistan
Tel: +92 51 819781-6
Fax: +92 51 826362
E-mail: ansar@worldbank.org

WEDC
Loughborough University
Loughborough
Leicestershire
LE11 3TU
United Kingdom
Tel: +44 (0) 1509 222885
Fax: +44 (0) 1509 211079
E-mail: WEDC@lboro.ac.uk
Website: www.lboro.ac.uk/departments/cv/wedc

Community waste collection

In many urban areas of Bangladesh, especially in Dhaka, the capital, there are serious problems with the disposal of household rubbish. The local authority regards refuse collection as a low priority. The collections made by the large trucks of Dhaka City Corporation are restricted to the main roads because they are unable to manoeuvre in the narrower streets. The authorities have had difficulty in coping with the quantity of rubbish produced, so the streets of Dhaka have gradually acquired piles of waste, sometimes left for months before they are cleared away.

After working abroad for some years, Mahbob Ahsan Khurram returned to Dhaka in 1987 and was shocked to see how the area in which he used to live had filled up with uncollected garbage. The air had an unpleasant stench generated by the rotting waste. In the years that he had been away, the population had increased and blocks of flats had been built to accommodate all the new residents who in turn were creating greater amounts of rubbish.

Mahbob decided to take action to clear up the rubbish in the streets and drew up a plan for community-based household rubbish collection. He set up a trial scheme in his own district of Kalabagan, which would receive payment only if it was successful and sustainable.

Collection of the waste

Waste is collected from the households in the district of Kalabagan every day. It is loaded on to rickshaws that have been converted to carry out the task of waste collection. The waste is then sorted for recycling and taken away to where the Dhaka City Corporation trucks are able to collect it.

The households in the district are made aware of the rickshaw's presence by the distinctive sound of the sweeper blowing his horn. When the horn is sounded, people bring their rubbish to the rickshaw. For the people who are unable to make their way to the rickshaw, a small additional charge is made for the sweeper to come to the house to collect the bags of rubbish. Alternatively, apartment dwellers who are reluctant to make the trip down to the street to dispose of their rubbish but do not want to pay the extra fee for the sweeper to collect it from their door have come up with the simple solution of lowering their rubbish down on a string, from their balcony or window, to the rickshaw in the street.

The working teams on the rickshaws

Mahbob has two rickshaw vehicles and he employs two municipal sweepers and three other people (usually relatives of the sweepers) per rickshaw. They earn two to three times as much as a part-time Dhaka City Corporation street sweeper.

In addition to the working teams there is also a standby sweeper and someone else is employed to collect the money. A rickshaw team can collect from between 100 and 120 houses in one trip and will usually do three trips in one day.

Conversion of the rickshaws

The rickshaws had to be converted in order to be suitable for household rubbish collection. Pieces of 16-gauge sheet steel were welded together to make large boxes (183 × 91 × 107 cm) to hold the rubbish.

A rickshaw designed for the collection of waste

An initial investment is required to purchase the specially constructed rickshaws, equivalent to around twice the monthly wages bill. They need to be replaced regularly due to the harsh working conditions, although painting them regularly helps to prolong their life.

The only other tools required are spades and shovels that are used to unload the rubbish.

Recycling the waste

Extra income can be generated through recycling the waste once it has been collected. The waste is sorted, separating the plastic, paper, metal and glass from the rest of the waste, such as the green waste from the kitchen, and the materials are then sold on to recycling enterprises in the city.

A daily collection of rubbish is now established, covering 700 houses and 300 shops. The customers pay a small sum each month for the collection of their rubbish, although collections are also made from a number of households that do not pay the monthly charge, in the hope that they will join the scheme at a later date.

About 15 per cent of the scheme's users regularly default on their payments, but the scheme is self-supporting. The income from the rickshaws covers the wages of the workers and the equipment that needs to be purchased but, most importantly, it improves the standard of living for the residents in the area.

Benefits of the household waste collection scheme

Now that the area is clean, people are more reluctant to dump their rubbish in the streets and the residents actively discourage any careless dumping. Rainwater is free to drain away because there are no piles of rubbish blocking its path and so the streets no longer smell.

Support for the scheme came from the Chief Engineer of Dhaka who has also encouraged similar schemes for the collection of rubbish in other parts of the city. Mahbob's refuse collection enterprise has generated a lot of interest throughout Bangladesh and other parts of the world. In Dhaka, 29 other neighbourhoods have copied Mahbob's scheme, and another team is now operating in Kalabagan.

Mahbob recognizes that the schemes have to remain small – that is, no larger than about 1000 household collections – otherwise they will lose their effectiveness. Using local operators means that the residents know the people that they are dealing with and this makes the collection of money for the service easier.

Further information

UNDP/World Bank Water and Sanitation Programme RWSG-SA
Flat No. 01-01
Priyo Prangan
2 Paribagh
Dhaka
Bangladesh
Tel: +88 2 865241/504472/504249
Fax: +88 2 865351

Automated recycling of drinks containers

Globally, there are approximately 700 billion drink containers currently in circulation. In the United States alone, more than 100 billion units of aluminium cans and 15 billion units of lightweight, non-breakable plastic bottles are produced each year. As the demand for drinks increases, the need to find an effective way of dealing with the mountains of plastic cartons, cans and bottles becomes more urgent.

Without organized systems for collecting back or incinerating used containers, the containers will ultimately end up in landfill sites or littering the environment. The sheer magnitude of raw materials and energy needed to produce these containers also represents an increasing threat to the world's supply of natural resources.

Reducing litter and landfill

Public legislation plays an important role in encouraging reuse and recycling. Public services such as kerbside collection of recyclable waste, central waste collection centres and energy-efficient incinerators are a step in positive direction. Better still, mandatory deposit/

refund systems have proven that consumers are positively influenced by a monetary incentive to return empty containers for recycling.

Reverse vending machines are a convenient way for consumers to return their empty beverage containers. They can be installed in practical locations, such as supermarkets and collection centres, and since they automate the handling process, they can be used at any time of the day, at the customer's convenience. They also help to protect the environment by reducing litter and conserving valuable natural resources.

The Norwegian company, TOMRA, manufactures reverse vending machines, which are used to collect and identify used beverage containers for recycling or reuse from consumers.

Cost-efficiency

When TOMRA pioneered the reverse vending machine business in 1972, the initial idea was to help supermarkets handle the return of empty beverage containers in a more cost-efficient manner, which would in turn allow them to provide better service to their customers. Over the years, the original concept has extended to encompass not only the efficient collection of used beverage containers from consumers, but also the entire chain of events whereby the containers end up at their final destination (at a recycling or bottling facility). This approach offers benefits to all parties involved in the recycling process:

- The consumer is rewarded with money or some other kind of incentive.
- The supermarket saves time and money by having a reliable system handle the logistics of container return for them.
- The bottlers have an independent third party taking care of the complex logistics, transport and accounting involved in public or private deposit systems.
- The recyclers receive the waste material in the right form, at the right place, at the right time.

Packaging recovery

TOMRA designs and operates systems for recovering packaging for reuse and recycling. It has developed a comprehensive series of products to cater for the different needs of each supermarket and collection centre. Reverse vending machines reduce the need for manual labour and time-consuming inventory and cash control.

Pre-sorting of glass and plastic bottles reduces handling time
Source: TOMRA

Non-refillable containers

Non-refillable containers (also known as 'one-way' containers) are not normally returned to the bottler for reuse. They are usually recyclable, however, and can be converted into other useful materials. For example, plastic bottles can be recycled into textiles and carpeting, glass bottles can be recycled into new glass bottles or insulation material, and aluminium cans can be recycled into car parts or new cans.

The company has devised a range of machines for handling non-refillable containers. A high-speed barcode reader ensures fast and reliable recognition. Once the container is correctly recognized, it is compacted, crushed or shredded, which reduces storage volume and makes for easier, more cost-efficient transportation.

Since contamination can reduce the value of the recyclable material, the machines can separate coloured glass or plastic into different storage tubs, separate aluminium and steel cans, and ensure that containers with excessive or unwanted residue are rejected.

Refillable containers

Refillable containers are normally returned to the bottler for washing and reuse. Refillable PET (a type of plastic) bottles can be reused up to 20 times, provided that they are not contaminated or damaged. Refillable glass bottles can be reused even more times.

Benefits to the consumer

- A faster, more convenient way to dispose of empty beverage containers.
- A more reliable method of reclaiming deposits, as the machine will always identify the container correctly and thereby issue the correct refund.
- The opportunity to experience new promotions on or around the reverse vending machine.

Benefits to the retailer

- Automated collection and sorting of returned containers saves time and money.
- Sanitation problems are reduced, since containers are sorted and stored in a more orderly fashion.
- One-way containers are crushed, compacted or shredded, which not only reduces the need for storage space, but also eliminates the risk of fraudulent multiple redemption of containers.
- Accurate container identification ensures that the correct refund is paid out.
- Reliable accounting features ensure that the system cannot be cheated.
- The reverse vending machines and store design concepts offer the retailer's customers a faster, more convenient and reliable way to dispose of their empties. This helps give an overall good impression of the store and can lead to increased store traffic.
- Value-added consumer services, such as couponing and fund-raising, make returning empties a far more enjoyable and rewarding experience and help give the store a competitive advantage.

Further information

Caroline Quinn
TOMRA Europe AS
Marketing Communications Department
Drengsrudhagen 2, N-1385 Asker
PO Box 278, N-1372 Asker
Norway
Tel: +47 66 79 92 03
Fax: +47 66 79 92 30
E-mail: caroline.quinn@tomra.no
Website: www.tomra.no

Cash from cans

The beaches of Montevideo, the capital of Uruguay, are filled with sunbathers. In the past, sunbathers have failed to dispose of their empty drink containers properly, turning the beaches into unsightly places, as well as potentially hazardous areas for beach users. Uruguay is not a country recognized for its interest in green initiatives but, with the number of aluminium cans being imported on the increase, people needed to be encouraged to collect and recycle their used cans.

The aluminium drinks can is one of the few objects that can be recycled so as to produce exactly the same object all over again. Furthermore, since remelting aluminium requires only 5 per cent of the energy used in the original smelting process, recycling the cans makes a positive contribution to energy conservation and ensures an economic source of metal for the future.

Prolata

Prolata is an education initiative that has been established to collect waste and tidy up the beaches. The initial campaign involved the setting up of an efficient collection, cleaning and storage operation

Aluminium can be recycled successfully to conserve energy as well as metal

which allows the aluminium drinking cans to be recycled. At the same time, Prolata has been pioneering a major public educational programme based on the promotion of a new culture in Uruguay – reducing, repairing, reusing, recharging, restoring and recycling.

Alcan Aluminium Limited

The multinational company, Alcan Aluminium Limited, is recognized as one of the world's largest producers of aluminium and a leading manufacturer of aluminium-based products. Its network of operations is carried out across six continents. In 1990, Alcan was responsible for recycling 325 000 tonnes of aluminium and its recycling capacity has continued to increase since then. In the United States alone, Alcan annually recycles about 11 billion beverage cans.

Alcan is the main manufacturer supplying aluminium cans for the Uruguayan market even though it does not produce cans in either Uruguay or Argentina. The nearest manufacturing plant to Montevideo is in Sao Paulo, in Brazil. The NGO that runs Prolata negotiated a special deal with Alcan, who agreed to take the aluminium cans back to Sao Paulo from Uruguay, and pay for the transport, provided that enough cans were collected to make the scheme financially viable.

Prolata have to collect 20 tonnes of cleaned cans (equivalent to about 62 000 cans) which are then roughly squashed in a homemade machine prior to being crushed into 7 kilogram blocks at the Alcan depot in Uruguay and transported back to Brazil. It is not possible to melt the cans in Uruguay, but Alcan helps with the collection and compression of the cans before sending them on to the melting plant in Brazil.

Recycling bins

Prolata's first task was to set up 200 collection points all over the city of Montevideo. With the help of Montevideo's Town Hall rubbish collection department and the schools and community centres, bright blue bins, supplied by Alcan, were put in various places around the city. Collection points included the beaches, service stations, schools and general stores. Prolata then raised funds for a van that would carry out regular collections of the large plastic bins, replacing the full ones with empty ones.

The Tacuri Association

With the logistics in place, Prolata made an alliance with the Tacuri Association which creates employment opportunities for young people from deprived areas. The young recruits are responsible for going out, along the beaches of Montevideo, and encouraging all the beach users to throw their empty cans into the designated bins.

Crushed aluminium cans awaiting re-export

Income generation

Prolata sells the aluminium cans to the Alcan factory and the income generated goes towards helping schools and people in need.

The project combines a way of keeping the streets and beaches cleaner with an educational effort to teach the people of Uruguay about recycling. Running parallel to the actual collection of cans, an educational programme takes the recycling concept into school classrooms and adult workshops.

Further information

Prolata
Maldonado 1792
11200 Montevideo
Uruguay
(PO Box 6149)
Tel: +598 2 487 999
Fax: +598 2 420 228

Alcan Aluminium Limited
Public Relations Group
1188 Sherbrooke Street West
Montreal
Quebec
Canada
H3A 3G2

Accessories from inner tubes

Unlike developing countries, where there is a culture of reuse and recycling, in the United Kingdom – as in many industrialized countries – potentially reusable items are often burned or buried in landfill sites. Old inner tubes from the wheels of vehicles ranging from tractors to trucks have many uses, swimming aids and water containers being two simple examples, but in the United Kingdom most old inner tubes are discarded. There have been moves, however, to find new uses for these and other waste products.

Advantages of reclaiming and recovering rubber

Rubber recovery can be a difficult process. There are many reasons, however, why rubber should be reclaimed or recovered:

- Recovered rubber can cost half that of natural or synthetic rubber.
- Recovered rubber has some properties that are better than those of virgin rubber.
- Producing rubber from reclaim requires less energy in the total production process than does virgin material.
- Disposal of unwanted rubber products is often difficult.

- It conserves non-renewable petroleum products, which are used to produce synthetic rubbers.
- Recycling activities can generate employment.
- Many useful products are derived from reused tyres and other rubber products.
- If tyres are incinerated to reclaim embodied energy then they can yield substantial quantities of useful power. In Australia, some cement factories use waste tyres as a fuel source.

Rubber production

Rubber is produced from natural or synthetic sources. Natural rubber is obtained from the milky white fluid called latex, found in many plants; synthetic rubbers are produced from unsaturated hydrocarbons.

Natural rubber is extracted from the trees in the form of latex. The tree is 'tapped'; that is, a diagonal incision is made in the bark of the tree and as the latex exudes from the cut it is collected in a small cup. The average annual yield is approximately 2.5 kg per tree or 450 kg per hectare, although special high-yield trees can yield as much as 3000 kg per hectare each year.

The gathered latex is strained, diluted with water and treated with acid to cause the suspended rubber particles within the latex to coagulate. After being pressed between rollers to form thin sheets, the rubber is air- (or smoke-) dried and is then ready for shipment.

Crude latex is made up of a large number of very long, flexible, molecular chains. If these chains are linked together to prevent the molecules moving apart, then the rubber takes on its characteristic elastic quality. This linking process is carried out by heating the latex with sulphur (other vulcanizing agents such as selenium and tellurium are occasionally used, but sulphur is the most common). There are two common vulcanizing processes:

- *Pressure vulcanization*: This process involves heating the rubber with sulphur under pressure at a temperature of 150°C. Many articles are vulcanized in moulds that are compressed by a hydraulic press.
- *Free vulcanization*: Used where pressure vulcanization is not possible, such as with continuous, extruded products, it is carried out by applying steam or hot air. Certain types of garden hose, for example, are coated with lead, and are vulcanized by passing high-pressure steam through the opening in the hose.

The process of vulcanization gives increased strength, elasticity and resistance to changes in temperature. It renders rubber impermeable to gases and resistant to heat, electricity, chemical action and abrasion. Vulcanized rubber also exhibits frictional properties that make it appropriate for pneumatic tyre application.

The raw materials that make up tyres are natural and synthetic rubbers, carbon, nylon or polyester cord, sulphur, resins and oil. During the tyre-making process, these are virtually vulcanized into one compound that is not easily broken down.

The major uses of vulcanized rubber are for vehicle tyres and conveyor belts, shock absorbers and anti-vibration mountings, pipes and hoses. It also serves some other specialist applications, such as in pump housings and pipes for handling of abrasive sludges, power transmission belting, diving gear and water-lubricated bearings.

Recycling tyres

The rubber used in tyres is a relatively easy material to reform by hand. It behaves in a similar manner to leather and has, in fact, replaced leather for a number of applications. The tools required for making products directly from tyre rubber are not expensive and are few in number, for example, shears, knives, tongs and hammers, as well as a wide range of improvised tools for specialized applications. Shoes, sandals, buckets, motor vehicle parts, doormats, water containers, pots, plant pots, dustbins and bicycle pedals are among the products manufactured from old tyres.

Many garages pay firms to remove old inner tubes and, in the United Kingdom, 70 per cent of inner tubes are burnt or buried in landfill sites. Inner Tube Ltd is a company that produces eco-friendly products by turning recycled inner tubes into mirrors, cushions, coasters, bikinis, handbags, rucksacks, and accessories such as filofax and mobile phone covers. As they are made of rubber, these products are hard-wearing and waterproof.

Making a bag uses nine inner tubes and takes about eight hours to complete. The factory produces about 100 handmade bags per day. The bags are designed using brown paper sewn together to provide a template for the rubber. Inner tubes from cars and lorries can be used although tractor inner tubes are the best because they are the largest and easiest to sew. The straps of the bags are made from discarded seat belts.

Whether rubber tyres are reused, reprocessed or handcrafted into

Rucksacks made from old inner tubes
© Inner Tube Ltd

new products, the end result is that there is less waste and less environmental degradation.

Further information

Inner Tube Ltd
218 Fratton Road
Portsmouth PO1 5HH
United Kingdom
Tel: +44 (0) 2392 433433
Fax: +44 (0)2392 433434
E-mail: info@innertube.co.uk
Website: www.innertube.co.uk

Other useful websites:

www.itra.com/corporate/recycling/trrac.htm
International Tire and Rubber Association (ITRA) Home Page. A wealth of information on recycling of tyres and associated topics.

www.wrf.org.uk

Website of the World Resource Foundation

www.rapra.net

Website of RAPRA

www.usrubber.com/
A commercial website with an interesting range of products from recycled rubber.

From rags to handmade paper

Paper production requires a supply of raw materials and, as deforestation becomes an issue of international concern, environmentally sound methods that move away from large-scale wood-based paper production are becoming more popular.

The handmade paper industry in India offers considerable potential to meet the country's increasing demand for paper products in an environmentally sound way. The main raw materials used in handmade paper are cotton rags and waste paper, both rich in cellulose – an essential ingredient for papermaking. Delhi is an ideal location to find these materials because it has a thriving rag trade which provides an abundance of old cloth and vast amounts of used paper.

Handmade paper production has low capital investment, thereby promoting local entrepreneurship; it can be established in decentralized and rural areas; it generates more local employment; it is an environmentally sound technology, depleting fewer resources and causing less pollution than paper mill factories; and it can produce certain specialized varieties of paper, for example, watermark, filter paper and drawing sheets. The cost difference between handmade paper and millmade paper is marginal.

Rag chopping: the material is chopped into uniform-sized pieces

Sustainable livelihoods

Handmade paper production is an effective means of creating sustainable livelihoods in rural areas. The workers employed in papermaking plants do not usually require previous technical experience or knowledge. This enables the local population, especially unskilled women, to be trained to work in these plants.

Established in 1988, TARA (Technology and Action for Rural Advancement) has become a major manufacturer of handmade paper. It is the production and marketing wing of Development Alternatives, an international network dedicated to sustainable development, and operates on four basic principles. It aims to create new, local jobs – particularly for unskilled women – and currently employs 35 women and seven men. It makes products to fulfil basic needs, for example, paper. It conserves scarce resources – for example, wood and water – through using alternative materials and recycling. It minimizes pollution.

Production of handmade paper

The basic principles of handmade paper production are quite similar to those in large mills.

1. *Sorting and dusting*: The raw material is sorted manually to remove buttons, plastic, synthetic fibres and other foreign materials. It is also given a vigorous shake to remove the dust and dirt.
2. *Rag chopping*: The sorted material is chopped into small, uniform-sized pieces.
3. *Beating*: The raw material is mixed with water and inert chemicals and beaten in a Hollander beater. This is a U-shaped trough, with a drum, on the outer side of which are iron blades which cut the raw material to make the pulp. There is also a washing drum which cleans the pulp and removes the dirty water. The quality of the paper to be made determines the consistency of the pulp.
4. There are two methods of *sheet formation*:
- *Dipping* (for fine/thin paper): The pulp is diluted with water and put into a masonry trough or vat. The lifting mould (mesh on a wooden frame) is then dipped into the trough, shaken evenly and lifted out with the pulp on it. The consistency of the pulp in the tank should be kept constant.
- *Lifting* (suitable for all paper and card): A fixed measure of the pulp is poured evenly on to a mould, which is clamped between two wooden deckles (frames) in a water tank and dipped. The mould is then raised, using a lever mechanism, to drain the excess water.
5. *Couching*: After the sheet formation is completed, the wet paper is transferred on to a cloth/felt sheet and a stack of interleaved sheets is built up.
6. *Pressing*: A hydraulic press is used to remove the excess water from the sheets. Pressing reduces the bulkiness of the paper, i.e. the sheets become more compact. This improves the physical properties of the paper and facilitates drying.

7. *Drying*: After pressing, between 50 and 65 per cent of moisture remains in the sheets. The sheets may be dried by hanging them in open areas of sunlight to remove the rest of the moisture. Solar dryers can speed up this process and reduce the amount of space needed. Coloured paper is sometimes dried in the shade to avoid the bleaching effect of the sun.

8. *Cleaning and sizing*: Small dirt particles and other foreign matter are removed manually with a sharp instrument. The cleaned sheets are given a coating with starch to improve the physical properties of the paper and prevent feathering. This is called sizing and can be done manually with a brush or by the dipping method, where the sheets are immersed in a tub of size.

Paper sizing

9. *Calendering*: The sheets are placed between metallic plates and passed through spring-loaded rollers in a calendering machine. This smoothes the paper and enhances the gloss.

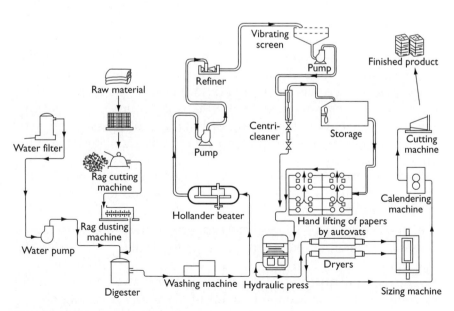
The papermaking process

10. *Cutting*: The sheets are neatly trimmed to the required size using a cutting machine.

The output can be increased considerably by implementing a few modifications to the manufacturing process described. This process (shown in the diagram) is used in India.

Further information

Development Alternatives
B-32 TARA Crescent
Qutab Institutional Area
New Delhi – 110016
India
Tel: +91 11 696 7938
Fax: +91 11 686 6031
Website: www.devalt.com

Paper from algae and other wastes

The papermaking industry is widely regarded as one that has paid little attention to the environmental effects of its activities. It is seen as using up natural resources, such as land, trees, water and energy, to make a final product that has a very short useful life. After only one use, most paper becomes a waste product that presents a nuisance and contributes to the overall waste disposal problem, especially in urban areas. Since the 1970s, public opinion in many parts of the world has put increasing pressure on the paper companies to improve their attitude to environmental issues, but the industry as a whole still consumes large amounts of timber as its main raw material and is a major user of water.

Recycled paper

Public pressure and new laws have encouraged the industry to recycle more waste paper, to reduce the demand for timber and, worldwide, large amounts of waste are now reused. However, although paper recycling is a positive step, it is not the perfect answer, either from the environmental or the economic point of view. The recycling process requires the removal of printing ink, which generates a polluting residue. Also, because repulping damages the fibres which make up the paper, the final product is always considered to be inferior to 'new' paper – and so it commands a lower price. The alternative is to replace cellulose fibre from trees with other materials.

The Italian paper company, *Cartiera Favini*, has taken more radical steps to improve the environmental situation in its own production. The company, in the Veneto region of north-east Italy, was founded in 1736 when a windmill was converted into a paper factory. It used textile waste as its raw material, producing 20 kg of paper per hour. Since then, the company has grown and now produces some 18 000 kg of paper per hour. The company, which is relatively small by paper-industry standards, is not typical of the industry as a whole because it concentrates on 'speciality papers', operating in a niche market for products that have a high quality and command high prices and for which the demand is small.

In 1990, the senior management recognized the value to the company's sales and image arising from an environmentally sound production policy and took the bold decision to find ways to reduce its use of natural resources. By pioneering the use of new raw materials to substitute for the conventional papermaking materials, it has demonstrated that unconventional materials can produce high-quality paper products without serious financial implications. In 1991, the company embarked on its *EcoFavini* ecological production programme which began to introduce a range of products with improved environmental benefits. It now has a policy of finding innovative ways of using the wastes of other industries as its own raw materials.

Algae paper

In the early 1990s, serious pollution in the Adriatic sea resulted in the Venice Lagoon becoming infested with thousands of tonnes of algae, which caused serious problems by reducing the natural oxygen level in the water, and this in turn killed fish and caused bad smells. In order to combat these problems, the algae are collected from the lagoon and brought ashore for disposal. The 50 000 tonnes of algae collected each year is the papermaking equivalent of 30 000 tonnes of trees. The research and development team at *Cartiera Favini* succeeded in finding a way of processing some of this waste so that it can be used as a substitute for some of the cellulose fibre in its papers.

The algae must be dried immediately in order to ensure that it does not begin to rot and smell and then it is finely ground into a powder or 'flour'. No chemicals are used in the process. The flour is then added to the paper pulp, as a replacement for some wood fibre, and the pulp is used in conventional machinery to make high-quality paper. There is a significant saving in fuel use because processing the algae needs only half as much energy as wood cellulose processing. The effect in the *Favini* mill is that, in each working day, as well as 20 trees, 40 tanks of fuel are saved. The product, like all the others in the company's range, is recyclable and free of both acid and chlorine.

In the very early stages, the costs of production were three times those of conventionally produced papers with similar qualities but now the costs are close to those of traditional paper. It is expected that the costs will eventually be very much lower.

Sugar paper

Italy grows large quantities of sugar beet and, after the sugar has been extracted, the remaining pulp presents a disposal problem to the sugar industry. At the *Favini* paper mill, this waste has been successfully used to develop a product which they call 'sugar paper'. This represents an important development, because it takes the waste from one industry and uses it as a raw material in another. There are, therefore, two good environmental effects – the papermaker is able to save trees and the sugar industry does not have to find a way of disposing of a possible pollutant.

Other papers

In a similar way to sugar paper, the company has also succeeded in using the fibrous wastes from maize processing – stalk, leaves, cob and bran – to substitute for some of the wood cellulose used in paper.

Both sugar and maize residues contain some fibres that have similarities to wood cellulose, but the *Favini* company has found ways of using other residues that have little or no fibre content, such as the residual pulp of citrus fruit pressing and grape pressing. In these cases, the pulp is dried and very finely ground to make 'flour'. The flour can then be used to replace an equal amount of wood cellulose and mineral fillers in the paper process. The company gives the products names that come from the original waste, such as 'Orange paper', 'Lemon paper' and 'Wine paper', which the consumer can identify easily as showing that these papers are environmentally sound.

Smog paper

The company's most startling recycling process is used in what they call 'Smog paper'. The factory's furnaces, like all fossil fuel burners, produce effluent gases which are universally recognized as harmful pollutants. The *Favini* company found a way to collect these gases and 'fix' (neutralize) them on alkaline industrial residuals such as the ceramic 'slips'. The pilot plant is able to treat 1000 m^3/h of smoke (combustion gases), thus eliminating the pollution, which is equivalent to that produced by 30 cars travelling at 50 km/h.

The neutralized gases are transformed into a harmless powder called 'smog flour'. This neutral white flour is used as inorganic filler in place of calcium carbonate, to produce a paper with a recognizable name.

Sludge

All papermaking processes leave a residue of fine materials which have not been caught up in the paper as it is formed in the machine. The materials are suspended in water and carried away for disposal. In the *Favini* factory, the resulting scum on the water in the settling tanks is skimmed off during the water cleaning process. The water is reused and the scum is fed back into the pulp stage for further savings. The reuse of the scum materials effectively increases the efficiency of the production process by reducing raw material costs. The recycling of water has had a dramatic effect on the company's water consumption. Over a period of eight years, *Favini* doubled its production and reduced the amount of water it took from the mains supply from 2 million to 569 000 cubic metres per year.

Environment versus profitability?

The company's policy of following an environmentally friendly policy does not seem to have interfered with its ability to perform well financially. The ecological products have generated great interest in the news media, both within Italy and internationally, and have won several awards for the company. The result has been to stimulate the market for all the company's products due to the improvement in its public image. Compared with other Italian paper companies quoted on the Milan stock exchange, *Favini's* income performance has been significantly better than the average.

Further information

Geopolimeri Srl – Gruppo Favini
Via Cartiera, 21 – I-36028 Rossano Veneto
(Vicenza)
Italy
Tel: +39 424 547770 / 84722
Fax: +39 424 84509
E-mail: geopol@tin.it

Rehabilitating water hyacinth

Covering 69 000 square kilometres, Lake Victoria is Africa's largest inland body of water. It borders the nations of Uganda, Kenya and Tanzania. About 30 per cent of Uganda is covered by water and ultimately most people depend upon the lake for their livelihood. More and more of Lake Victoria is disappearing beneath water hyacinth and it is estimated that 100 square kilometres of the lake's surface is covered by the weed.

Water hyacinth reduces light and oxygen and threatens biodiversity

The high doubling rate of the water hyacinth – under ideal conditions, quantities can double every two weeks – means that it has become a major environmental nuisance. It grows in mats up to two metres thick which can reduce light and oxygen, change water chemistry, affect flora and fauna, and cause significant increases in water loss. It also causes practical problems for marine transportation and fishing and it is now considered a serious threat to biodiversity.

Extensive amounts of water hyacinth have clogged the Port Bell Pier, which is the main outlet from Uganda on Lake Victoria. This is hampering the fishing canoes used in local water transport systems, fishing, navigation, and the incomes and livelihoods of the communities dependent on the lake. There are now more inaccessible shorelines, more unreachable fishing areas and fewer fish because of the increased plant coverage reducing the oxygen in the lake.

Rehabilitation of prisoners

The Murchison Bay Reserve, situated 10 km outside Kampala and lying adjacent to Port Bell Pier, houses a group of prisons which hold 60 per cent of Uganda's total prison population. The Luzira prisons are mainly inhabited by capital offenders serving at least five year sentences and often life imprisonment. As a form of rehabilitation, there are several training programmes that some of the inmates can join, for example, carpentry, tailoring and leather works.

The initial phase of eradicating the water hyacinth from Lake Victoria focused on simply harvesting it and using it as animal feed, but environmentally friendly solutions for the raw material have been sought with a view to improving prison welfare.

In Bangladesh, the Mennonite Central Committee (MCC) carried out experiments using water hyacinth to make rope, which could then be woven around bamboo frames to make furniture. MCC now assists furniture makers in Bangladesh with designs and helps them to market their environmentally friendly products.

Alternative uses of water hyacinth

A group of Mennonites in Uganda passed on the information, about water hyacinth being used in Bangladesh to make crafts and furniture, to the Upper Prison in Kampala. The Upper Prison holds about 2000 maximum security prisoners and it has taken them two years to learn the techniques that transform the hyacinth stems into a range of saleable products.

The weed is processed to make ropes that can be used to make furniture and handicrafts for selling, such as shopping baskets, as well as sleeping mats for the prisoners. MCC also gave the prison instructions for water hyacinth papermaking, which uses 50 per cent hyacinth stem and 50 per cent waste paper, although this was not as popular among the prisoners.

How to process water hyacinth

- Collect the water hyacinth from the lake.
- Transport the weed from the lake to the prison – a distance of 2.5 km.
- Remove the roots and leaves – these can be used as raw materials to sustain a biogas plant.
- Split the stems lengthways and allow to dry in the sun for a day.
- Scrape out the inner pith of the stems with a knife.
- Allow the stems to dry in the sun for a further three days.
- Soak in a solution of sodium metabisulphite or caustic soda and water for one hour (to preserve the fibre and stop the rope from rotting) – these chemicals are totally used during processing leaving a chemical-free residue at the end of the process.
- Dry in the sun for 5 to 6 hours.
- Cut the stems lengthways into strips (the width depends on the diameter of the rope required).
- Prepare rope by braiding three strips of dried stem.
- Optional – the rope can be boiled in dye at this stage.
- Cut off loose strands from the rope.
- Weave furniture or handicrafts.

Transporting the water hyacinth

Water hyacinth is 95 to 97 per cent water and it is bulky and heavy. Once it has been dried out in the sun, an entire truckload of water hyacinth reduces to a sack of stems weighing only 10 kg. This means that to make enough rope for one armchair or two dining chairs a full truckload is needed. Transportation of the harvested weed is costly due to its extremely high water content, although chopping can reduce both the volume and the water content.

Benefits and opportunities

The water hyacinth furniture trade is labour intensive, which reduces the idleness of the prisoners and equips them with skills they will be able to use after they are released. It also improves their self-worth and dignity. Each of the inmates earns 2 per cent of the profits on every item produced. This provides them with an opportunity to generate income while in prison so that they can purchase small commodities like cigarettes and toothpaste.

The project is financed by a grant from the United Nations Development Programme (UNDP) through the Small Grants Programme of the Global Environment Facility. It is targeting 500 convicted male and female long-term prisoners. The inmates from Upper Prison are training women from the neighbouring jail in the techniques of manufacturing water hyacinth furniture and craft products. The authorities believe that as well as the financial benefits derived from the water hyacinth processing, the activity is helping prisoner rehabilitation.

The prisoners are trained in weaving and handicraft skills by external experts and environmental bodies. The training covers environmental education and biodiversity conservation issues; weaving; crafts making and design; papermaking; and disposal methods. An effective training programme must also involve training trainers so that the project remains sustainable in the prison. Eventually, the project is intended to be self-sustaining through the sale of products.

The East African Wild Life Society (EAWLS) is helping with the marketing of the water hyacinth products from the prison. Currently the outlet channels are the prison showroom, exhibitions in the Sheraton Hotel in Kampala, displays at the Parliamentary Buildings and CNN adverts. Individual orders can also be taken directly.

It is hoped that the removal of the water hyacinth from the lake will improve the terrestrial life, maintain the food source, transport and other commercial activities, as well as improve the livelihoods of the communities dependent on the lake.

Further information

UNDP Resident Representative
PO Box 7184
Kampala
Republic of Uganda
Fax: +256 41 244801
E-mail: fo.uga@undp.org
Website: www.undp.org

Fuel from plastic waste

China's society has seen dramatic changes over the past few years. Until recently, China was a largely agrarian society, but it is making a rapid transition into a more urbanized and developed one, with consumer goods such as fast food and fast cars becoming more readily available. With a fifth of the world's population living in China, the quantity of processed and packaged goods that the country has the potential to consume is vast. As a result, waste, and specifically plastic waste, has become an enormous problem.

Beijing alone produces 120 000 tonnes of plastic waste each year and the infrastructure for waste management is straining to keep up with the increase. Landfill sites are at a premium. Recycling of the waste plastic is difficult to do once the various plastic materials have been mixed together because each plastic item needs to be identified and sorted before it can be recycled.

Seeking solutions

A group of Chinese industrialists and entrepreneurs came to the conclusion that the best solution to this problem would be to use the plastic waste as a resource to make fuel. This would not only reduce the large amounts of rubbish being produced but could also be used to provide much needed diesel fuel and petrol. The concept of obtaining liquid fuels from waste plastic was initiated in the early 1970s when oil prices suddenly increased dramatically. A great many countries, including Japan, Germany and the United Kingdom, conducted laboratory experiments and pilot schemes on liquid fuel production.

Experimental work into the conversion of plastic into fuel began in China and currently there are factories working to produce fuels using the techniques developed from these pilot schemes. Most of the manufacturing processes have been on a small scale and, in the past, they have been of a low technological standard with a bad smell and low stability. However, the process has been developed over the years and it is being refined continually to increase the general performance and process efficiency.

Recyclable plastics

Any old plastic products (including tyres) are suitable for use in the conversion process, although PVC needs to be treated with special attention using modified catalysts, etc. Typically, plastics are made from polymers such as polyethylene, polypropylene and polystyrene. These plastics are made from long hydrocarbon chains and have a very high energy content, and so they are ideal materials for fuel production.

The manufacturing process

The whole process takes about six hours but it is run on a continuous basis. All technological parameters, such as weight, temperature, pressure and flow rates, are automatically monitored and controlled to ensure that the process runs smoothly.

Unwanted gaseous emissions are filtered, and cooling water is circulated in a closed system so that pollutants are kept at an acceptable level. There is also a small amount of solid waste residue which needs to be disposed of at intervals during the process. The final products are stored in separate tanks where they are ready to be sold to transport companies who will use the fuel to power their fleets of trucks.

The use of a catalyst

Waste plastic is separated from the rest of the rubbish and bought from recycling centres and other local organizations. It is then taken to a conversion plant where the uncleaned waste plastic is crushed to pieces of 3–6 cm, before it is passed through a cyclone and a sieve to remove any dirt.

Once the dirt has been removed, the plastic is loaded, using a hydraulic feeder, into the cracking reactors where it is heated in the presence of a catalyst. The catalyst is important because it lowers the amount of energy that is required to break down the structure of the waste plastics.

As well as promoting the initial cracking of the polymers, the catalyst is used to promote the production of a heavier fuel suitable for the manufacture of diesel and petrol. Only a small amount of catalyst material is lost during the conversion process.

The collection of liquid fuel

The reaction chamber is specifically designed with a stirring device set at the bottom to prevent coking and to promote the removal of residue. From the initial cracking process, gases are driven off. As the plastics are reduced, the gases are collected and cooled, yielding liquid fuel. This liquid fuel or crude oil is a complex mixture that has

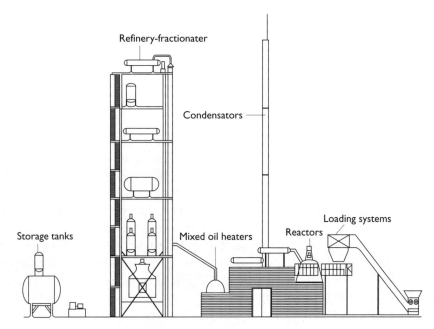

The conversion plant turns waste plastic into liquid fuel

to be separated in a fraction chamber to form petrol and diesel. The remaining incondensable gases pass through the top of the fraction chamber and are either burnt off in a flare stack or fed back to the initial stage of the process, where they are used as an additional fuel to heat the incoming plastic materials.

Heating sources

In China, coal is in abundant supply and it is cheap to buy. It is used as the main fuel to heat the plastic and catalyst reaction. Any alternative heating source could be used, depending on local circumstances. Designs exist for using oil-fired or gas-fired systems, even the diesel produced by the process could be fed back to heat the waste plastic.

Costs

At the moment in Beijing there is no systematic rubbish separation and collection programme that could deliver large quantities of waste plastic for processing. As a result, the waste is collected and sorted on a small scale in a number of different locations in the city. This has meant that transporting the waste is more expensive than is strictly necessary and the quality of the materials is extremely variable. The largest single cost in producing the fuel is the purchase price of the waste plastic from the recycling companies.

The existing factories in Beijing are too small to deal with the amount of waste plastic currently being produced so investment has been made for the construction of a large-capacity plant that will be able to cope with 60 000 tonnes of plastic annually.

'Best Practice'

The process has been praised as a 'Best Practice' by the Chinese State Environmental Protection Administration and has been profiled as a project with commercial potential by the National Science Foundation of China. There is a possibility that other chemical products could be obtained from this process. For example, the industrial solvent, toluene, could be manufactured, or xylene – which is used in the production of polyester. Ethylene, used for ripening fruit, could also be produced.

Further information

China Aerospace Great Wall Group
Research and Development Centre
No.22 Fucheng Road
Beijing 100036
P.R. China

Beijing Energy Efficiency Centre (BECon)
Zhansimen, Shahe
Changping, 102206
Beijing
P.R. China
Tel: +861 6973 2059/6973 5234/6973 3114
Fax: +861 6973 2059
E-mail: becon@public3.bta.net.cn
Website: www.gcinfo.com/becon/becon.html

Shipping containers as building materials

Almost every town and city in the developing areas of South Africa is surrounded by growing shanties and squatter areas. Indeed, in many cities the areas of informal housing far exceed the size of the formal city. The people in these informal settlements do not have the assets necessary to gain formal access to land, building materials and housing rights. Adequate shelter is a basic need for a growing number of people and it requires new solutions to address the problem.

Cape Town is a beautiful city, situated on Table Bay at the foot of Table Mountain. It is the seaport capital of Cape Province and, after Durban and Richard's Bay, it is the third busiest port in South Africa. There are many settlements and shanties growing around Cape Town where people are living without adequate housing and infrastructure. In a country where the majority of the population still lives in inadequate housing, any form of building using locally available materials, at costs lower than conventional building methods, is beneficial to the residents of the settlements.

Shipping containers

Shipping containers are designed to strict shipping standards for transporting goods on ocean crossings. Ocean Shipping Consultants estimate that the global fleet of containers numbers approximately 10 million. They are usually made from mild steel or aluminium and are normally available in two standard sizes – 20 feet long by 8 feet wide by 8 feet high (610 cm × 245 cm × 245 cm) or 40 feet long by 8 feet wide by 8 feet high (1220 cm × 245 cm × 245 cm). The containers are extremely strong and watertight, although they do have a limited life at sea which is, on average, between 10 and 12 years.

This means that shipping companies have a steady supply of old containers that they no longer require but which are still structurally sound. These old containers are often sold to brokers who will then resell them to companies to be used in a variety of different ways. They have been used as small electrical generator stations, water purification stations, telecommunication cells, laboratories, portable medical rooms, underwater repair chambers, shops, community halls and they can even be used to store class B explosives. However, most often they are used by builders and farmers as secure storage space.

Creative Solutions

The port at Cape Town has plenty of spare shipping containers and BP Community Affairs decided that they could be used as a starting point for improving the living conditions of the poorer people in the area. One shipping company, Safmarine, the South African Merchant Navy, donates all its redundant containers to Creative Solutions, a company in Cape Town that was set up to convert the shipping containers into livable spaces.

Creative Solutions has made almost 4000 conversions for local shops and hairdressers, nursery schools and community centres.

Converting the shipping containers

Conversions of the shipping containers are paid for by local businesses. The first step is to lay the foundations in exactly the same way that they would be laid for a conventional building. Before the containers are craned into position, any walls that need to be removed from them are cut so that they can be combined to produce bigger spaces or adjoining rooms. Once the containers are in their correct position they are bolted together. Using bolts to fix the containers together means that the option remains to move the containers at a later stage if required.

This small business and training centre at Ga-Rankuwa was built by Safmarine and Eskom
Source: Safmarine

Work is also carried out on the interior of the containers. The walls are clad with insulation and a wooden floor is laid down. Holes are cut into the new structure to create windows and doors for the converted containers. A conventional roof is constructed over the containers and bolted into place. Gutters are fitted and pathways are made to the doors before the whole building is painted.

Creative Solutions carry out as much conversion work as required, in order for the containers to be appropriately modified for their new use. The company is prepared to lay carpets, fit kitchens and they can even provide a lawn. It just depends on how much money is available and how much the local community want to do for themselves.

Lulhando Education Centre

Lulhando Education Centre was set up by Christine Mlumbi some ten years ago and now takes 238 children. She obtained the containers for the school about six years ago and is very happy with the improvement in the school building. Previously, there was one large room that was used by everybody, but now there are more rooms and the children can be separated into smaller groups with other children of their own age, which helps both the teachers and the pupils.

Fires within the squatter areas are quite common but now it is safer for the children in the school because the buildings will not burn down. Due to the insulation, the children can now stay cooler in the summer and warm in cold weather. This means that it is possible to reduce the fuel needs of the school and therefore reduce the running costs. The roof does not leak or make a noise when it rains.

Costs

Using the converted containers as material for buildings is about two-thirds cheaper than using the more traditional materials for constructing buildings. Furthermore, the converted containers provide a perfectly sound structure.

The ultimate beneficiaries of this project are the residents of Cape Town who now have secure and pleasant buildings to work and live in.

Further information

Peter Petersen
BP South Africa
PO Box 6006
Roggebaai
Cape Town
South Africa
Tel: +27 21 408 2190
Fax: +27 21 252139

Safmarine
Safmarine House
22 Riebeck Street
Cape Town
South Africa 8001
Tel: +27 21 408 6911
Fax: +27 21 408 6842
Website: www.cpt.safmarine.co.za/gauss/

Chapter Four
Transport for the future

Conventional transport is largely responsible for many types of pollution. Engine combustion discharges 80 per cent of the carbon monoxide, 70 per cent of the nitrogen oxides, 60 per cent of the hydrocarbons and 80 per cent of the lead oxides present in city air. Leaked oils and fuels infiltrate the soil and pollute rivers. Dark dusts and exhausts cause respiratory diseases and blacken historic monuments.

The carbon dioxide emissions from cars and other vehicles contribute significantly to the 'greenhouse effect', which is causing climate change on a global scale and aggravating problems such as water and food shortages. In rural areas of developing countries, where often no motorized transport is available, people are forced to walk ever further in search of water.

In this chapter the concept of 'intermediate means of transport' is introduced – vehicles such as wheelbarrows, bicycles and animal-drawn carts. Workshops in Kenya have equipped themselves to produce good quality wheels for these vehicles, and in the city of Kisumu, where pollution and congestion are a constant problem, bicycle taxis are proving a successful innovation.

The industrialized countries are just beginning to reduce carbon dioxide emissions, but need to achieve an 80 per cent reduction by 2050 if damaging climate change is to be halted. Those in the cities of developing countries seek a similar cut in pollution levels. This chapter presents several successful examples of emission reduction, including the provision of short-hire bicycles in the Netherlands, electric three-wheelers in Nepal, and electric hire-cars in France, as well as some promising designs for cars, buses and trucks that will be less damaging to the environment. In Nepal the electric vehicles have made such an impact that the diesel equivalent has been banned from the streets.

Finally, there is the road infrastructure itself, and soil stabilization is important in areas of extreme weather conditions and consequent erosion. In Nepal, bioengineering has been introduced to reduce the incidence of landslides and so make the roads safer to use.

Intermediate means of transport

In rural areas most people spend many hours each day transporting water, fuel and food to fulfil their basic needs. This restricts the time available for productive activities and is a significant constraint on social and economic development and hence on poverty alleviation. Lack of transport also reduces people's opportunities to earn a living from agriculture or other trades and may limit access to health, education and other essential services.

Bicycle taxis

Kisumu is one of Kenya's fastest growing urban centres. The current transport infrastructure is inadequate for the task of getting people from one place to the next and the roads are full of potholes. The streets are congested, with cars, buses and matatus (private minibuses) blocking them. Most commuters have no alternative other than to walk huge distances each day because it is quicker than using the transport systems available.

Standard bicycles are being modified and now serve as taxis. The drivers of the bicycle taxis are able to avoid the potholes and can weave through the traffic and the crowded streets. It is by far the quickest and most efficient form of transport. Bicycles can be ridden on a variety of terrain and their configuration makes them well suited for use on busy roads and narrow paths.

The bicycle provides the potential to increase the speed and range of travel considerably and the addition of simple, low-cost carrying devices gives an efficient means of transporting passengers and other goods. There are now 500 bicycle taxi drivers in Kisumu and, provided that they receive support from the town authorities, they could become real competition to the matatus and other motorized transport.

Bicycle taxis are now common on the streets of Kisumu

Intermediate means of transport

The access of rural people to motorized transport is very limited because of low income levels and poor road infrastructure and the introduction of lower-cost vehicles can bring major benefits. For instance, a wheelbarrow can carry three times as much as a person and may mean one trip for water per day instead of three; a bicycle can travel at roughly three times the speed of walking, providing much greater access to markets, job opportunities and essential services; and animal-drawn carts can carry up to one tonne, enabling produce to be moved rapidly from the fields, to reduce deterioration and wastage on the way to market.

These low-cost forms of transport are generally known as 'intermediate means of transport' (IMT) – i.e. intermediate between human and motorized transport – and are appropriate to rural areas in terms of meeting local transport needs, being affordable and incorporating suitable levels of technology.

The introduction of IMT may be particularly beneficial to women, as they are usually the main load carriers when other means of transport are not available. Health and safety are improved since, when there is no transport, people are often forced to carry loads which are far too heavy and in ways that may risk injury.

Production and supply of IMT

Some IMT, such as wheelbarrows and bicycles, are mass produced and imported into developing countries, while others, such as handcarts, bicycle trailers and animal-drawn carts are made locally. However, imported wheelbarrows of the type used in gardens are not suited to general rural needs of transporting quite heavy loads over long distances and more appropriate versions may be made locally. Also, bicycles are widely used for carrying loads and imported versions may be adapted to improve the transportation of goods and passengers.

A significant proportion of the local manufacture and adaptation of IMT is carried out in small and medium-sized enterprises (SMEs) located in the larger rural centres. Through their direct links with customers, these enterprises play an important role in supplying IMT to meet local demands and in providing repair services. They are also important to the development of rural economies.

Although enterprises are often quite innovative in adapting available means of transport to meet local needs, they generally need support in the form of information on appropriate designs, upgrading of manufacturing methods, and in promotion and marketing. A particular problem faced by most rural enterprises is access to good quality, low-cost wheel and axle assemblies. Scrap versions obtained from other vehicles are generally unreliable, while locally devised versions can be quite crude and inefficient. Improving the manufacture of wheel-axle assemblies is therefore often the principal need for increasing the supply of good quality IMT in rural areas. For example, it may be appropriate to promote the specialized manufacture of wheel-axle assemblies in one or two enterprises for supply to others that produce vehicles.

Wheel manufacturing technology

Access to wheel manufacturing technology enables workshops to set up their own facilities to produce a range of good quality wheels from standard steel sections. Wheels can be made to take bicycle, motorcycle, car and solid rubber tyres and to suit a range of low-speed vehicles such as handcarts, bicycle trailers, wheelbarrows and animal-drawn carts.

A system has been introduced by IT Transport which comprises a hand-operated bending device capable of forming good quality wheel rims from steel sections and an assembly jig (see photograph) to ensure that wheels are made to a consistent quality. The equipment

The assembly jig ensures that wheels are made to a consistent quality

can be made in a workshop that has competent metalworking skills and basic tools for cutting, welding, drilling and grinding steel. It can be readily adopted and used by other workshops with basic metalworking facilities to produce good quality wheels at low cost and on a small to medium scale of production. The wheel manufacturing equipment can be bought for about US$450 per set, plus freight charges. It may be possible to find a donor to supply the equipment to non-commercial workshops or training organizations, particularly where it is being introduced into a new area.

Making a wheel

The components of the rim are formed into rings in the bender and the wheel is then assembled and welded up in the jig. The steps are as follows:

1. The screw stop is set for the required diameter of the rim.
2. The rim section is cut to length and marked out in 2.5 cm (1 inch) steps.
3. Bending the rim in small steps produces a smooth, accurate circle.
4. The formed rim is clamped in an assembly jig to weld up the wheel.
5. The ends of the rim are cut to leave a 1 or 2 mm gap and the joint welded up.
6. The spokes are cut to size and welded into position.

Jua kali

Open air workshops producing stoves, cooking pots, watering cans and all kinds of ironmongery from scrap metal can be found all over Africa. In Kisumu, Kenya, the latest hardware is available from the open-air businesses – *jua kali*, which means hot sun. The artisan businessmen of the *jua kalis* have a thriving trade in transport technology and are finding novel solutions to the problem of getting people from home to work on the increasingly congested streets.

Further information

Ron Dennis
IT Transport Limited
The Old Power Station
Ardington
Near Wantage
Oxon OX12 8QJ
United Kingdom
Tel: +44 (0) 1235 833753 / 821366
Fax: +44 (0) 1235 832186
E-mail: ittthe@tranport.co.uk

Bicycle hire

In Amsterdam, the rise in traffic has intensified problems of accessibility and taken its toll on the quality of life. City dwellers and visitors are subjected to constantly increasing levels of air pollution and noise. Access to, and travel within, the city centre is problematic. The problem of finding convenient parking spaces is worsening. It is recognized that if this situation is to be improved it is vital to limit road traffic and improve public transport amenities.

Individual modes of transport have always been popular for travelling short distances, and the bicycle is an appealing, enjoyable and environmentally friendly alternative to trams, buses and the metro. According to research findings, the use of bicycles would intensify if availability increased, and that in turn would lower the statistical chances of bicycle theft.

The 'white bicycle' scheme

A hire scheme has been introduced in Amsterdam that serves primarily as the source of a convenient means of transport to and from tram and bus stops and railway stations. The scheme also has the potential to reduce pressure on the environment by encouraging people to use bicycles instead of other means of transport, and this could ultimately save the government money as fewer investments would be needed to improve the road infrastructure.

A total of 750 white bicycles and 45 bicycle depots have been introduced into Amsterdam's city centre and the surrounding districts to try to alleviate the traffic problems. This means that tourists and residents have a permanent means of transport within walking distance of almost any point in central Amsterdam. The chances are that there will always be a bicycle available in the neighbourhood bicycle depot and an empty space in the depot close to the cyclist's destination.

The white bicycle system works as follows: cyclists take a white bicycle from one depot and use it as transport to reach another location or transfer point. They return it to the depot that is nearest to their destination.

Features of the white bicycle

The white bicycle requires very little maintenance. The seat is easily adjusted with the touch of a button. To prevent punctures, the bicycle has massive tyres that operate like air tyres, but are foamed and leak-proof. It also features a battery and lights that turn on and

The distinctive design and heavy frame discourage theft of the white bicycles
Source: DEPO BV

off automatically whenever the weather conditions require, and a rack for luggage.

The bicycle is suitable for use by men and women alike. Each bicycle is fitted with an identification system, but does not have a locking device because it should be parked only within the specifically allocated depots, where the racks have built-in locks. The fairly heavy frame and unusual design are distinctive and this discourages theft of the bicycles.

The white bicycle is now being introduced as a facility at the bicycle shelters located at public transport terminals.

Smart cards

Around seven million people in the Netherlands have a 'Chipper' or similar multi-functional smart card. Anyone who owns a Chipper smart card can use it to remove a white bicycle and return it to another depot. This technology also enables user identification and ensures reasonable security. First-time users have to register their identity by inserting their card in the reader and keying in their code. From then on, the 'user profile' will be stored in the system and all the user has to do is insert the card to book and pay for a trip.

Bicycle depots

The depots are unsupervised areas, located approximately 300 metres apart and intended only for parking white bicycles. A depot consists of two racks, each with ten parking spaces, and a choice board. The bicycles are automatically locked on to the racks when they are placed into them. The choice board shows a map of the city with the locations of the depots.

The depots have a number of features, including a smart card reader for payment and user identification, a bicycle pump (for cyclists who have their own bicycles), a map to guide users in choosing their routes, a telephone and an Internet point. By using a smart card and clicking on the destination depot, the traveller books a journey to that particular depot and so a parking space is reserved. Only then can the traveller take a white bicycle out of the rack.

Redistribution and variable payments

When the traveller informs the selection stand of his/her intended destination depot, the system tries to book a parking space there, and refuses journeys to depots that are already full. It offers the nearest depot with spaces as a possible alternative.

By using variable payments, the system encourages journeys from full to emptier depots and discourages journeys from emptier to fuller depots. When the traveller chooses a destination depot, s/he is informed of the price which is then deducted from the smart card on booking.

In order to stop the depots becoming over-full or empty during the day, the redistribution of the bicycles is vital. Redistribution is accomplished by means of special bicycles which can carry six white bicycles at one time.

Further information
DEPO BV
Van Diemenstraat 76
1013 CN Amsterdam
The Netherlands
Tel: +31 20 625 0869
Fax: +31 20 622 6539
E-mail: depo@wxs.nl
Website: www.dds.nl

International Bicycle Fund: www.ibicycle.org/freebicycle.htm

Electric three-wheelers

The Kathmandu Valley is surrounded by hills on all sides and, because of its bowl-like topography, it is very susceptible to air pollution. Increasing urbanization, industrial activities and road traffic all contribute to air pollution in the valley.

The population of Kathmandu Valley was 577 246 in 1971 and has now increased to approximately one million. By 2015 it is expected to rise to around 1 800 000 as the population is increasing at an annual rate of about 5.7 per cent. As the urban population grows, the number of vehicles increases too, to meet the demands of the people.

In Kathmandu, pollution is most noticeable in the late spring and early summer months, at which time haze forms through a combination of dust, household smoke and car exhaust fumes. Increased emission of car exhaust fumes and other urban activities directly affect the mortality rates in a locality, at least among the high-risk roadside residents, shopkeepers and pedestrians. Petroleum-based fuels mainly generate hydrocarbons, carbon monoxide, carbon dioxide, oxides of nitrogen, lead, sulphur dioxide and other suspended particles which are hazardous to health and can cause chronic lung diseases such as bronchitis and asthma, as well as irritation of the respiratory tract, throat and eyes.

Simple observation reveals that much urban air pollution in Nepal, particularly in the Kathmandu Valley, is caused by vehicular emissions. The age and condition of vehicles appears to be a key factor affecting exhaust emission problems and the quality of fuel is also significant.

More than 38 per cent of the entire Nepalese transport fleet consists of three-wheelers, known as 'tempos', and motorcycles, both of which are mainly used in the commercial or public transport sector. These vehicles tend to be badly maintained and often use adulterated fuels. The use of low-quality fuel coupled with poor maintenance habits, which are partly attributed to the sparse availability and high cost of spare parts, result in the incomplete combustion of fuel. The excessive exhaust emissions which are produced contribute to a substantial increase in air pollution and consequently the deterioration of the atmosphere.

Electric vehicles

In order to try to reduce the problem of air pollution from vehicle exhaust emissions, two companies in Nepal have developed environmentally sound tempos which operate on batteries. These electric tempos cost the same to produce as those that run on diesel or petrol. Electric tempos use three batteries as their source of power, which need

NEVI's Safa tempo – a three-wheeled electric vehicle
Source: NEVI

TRANSPORT FOR THE FUTURE   CHAPTER FOUR

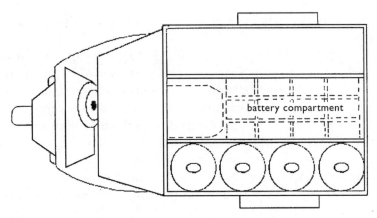

EVCO's electrically driven waste collection vehicle (viewed from above)
Source: EVCO

recharging every 60 kilometres. Each battery takes six hours to recharge and therefore almost a full day is required before the electric tempo can operate again once its batteries have run down. Despite this, an electric tempo has the advantage of having very few moving parts and an electric motor can last several years without any maintenance.

Nepal Electric Vehicle Industry, NEVI, which was established in March 1996, manufactures and operates electric vehicles (see photograph). It aims to promote electric vehicles, based on clean, renewable hydroelectric power as the appropriate means of transportation for Nepal and thereby reduce the country's atmospheric pollution as well as its dependence on imported fossil fuel.

The Electrical Vehicle Company, EVCO, manufactures three-wheeler electric vehicles. The chassis of the electric vehicles are manufactured by Scooter India Limited, Lucknow, and are claimed to be among the toughest ever made in India. All the electrical components used are manufactured in the UK or USA and the motor, specifically designed to carry heavy loads, is manufactured by the Prestolite Company.

Electrical equipment

The type of battery used in the electric vehicles is the Trojan T 105, which is made in the USA and is of the deep cycle lead-acid type. Ordinary vehicle batteries are quickly damaged if they are fully discharged very often but the deep cycle type is designed to withstand regular discharge. These batteries have a life span of about 700 to 750 complete charge and discharge cycles. Each battery produces 6 volts so that the 12 batteries provide a total of 72 volts. A fully charged set of batteries will drive the electric vehicle for a maximum of 60 kilometres.

In order to provide an operating voltage of 12 volts and, at the same time, provide the fairly high (15 amp) current needed by the motor, a DC to DC converter is used to reduce the storage voltage from 72 volts. The converter results in lower losses than would be experienced if resistors were used. A transistorized speed controller system is employed in order to minimize losses in this area and to extend the running time, thereby reducing operating costs.

An indicator lamp keeps the operator informed of the state of charge of the batteries as it is important that the batteries are recharged before they are more than 80 per cent discharged. The indicator lamp flashes when 70 per cent discharge is reached and then remains on when the battery level falls to the danger level.

All the lighting – front headlights, parking lights, front indicators, side indicators, brake lights and tail lights – run on the 12 volt circuit.

Specifications of electrical vehicles

Overall length	3200 mm
Overall height	1600 mm
Overall width	1430 mm
Minimum turning radius	7000 mm
Ground clearance	140 mm
Unloaded weight	650 kg
Payload	1000 kg
Rear wheel track	1250 mm
Rear axle weight	700 kg
Front axle weight	300 kg

Mechanical equipment

The electric vehicle needs only one gear because the maximum speed at which it can travel is 30 to 35 kilometres per hour. It has two braking devices: the handbrake, which stops only the rear wheels, and the footbrake, which controls all three wheels when the foot pedal is pressed.

The front suspension consists of a rocker arm which is assisted by two helical springs and telescopic hydraulic shock absorbers. The rear suspension works by using longitudinal springs, which are assisted by two telescopic shock absorbers.

The drive is to the rear wheels through the differential unit.

Servicing electric vehicles

The maintenance cost of electric vehicles is relatively low; however, the battery, used as a substitute for fuel, has to be replaced after 750 cycles of its charging and discharging process. This means that an electric vehicle can run for up to 45 000 kilometres before the battery needs to be renewed.

The following are required every 1500 kilometres:

- check gearbox oil
- check differential oil
- check brake fluid level in the reservoir
- check and, if necessary, adjust brake
- grease the brake pedal and the joints
- check steering box
- clean springs and grease all nipples using multi-purpose grease and suitable grease gun.

Every 2000 kilometres, the oil in the sump of the gearbox and the differential oil should be completely drained and replaced with new oil. The level of the oil can be checked with a dipstick and should be maintained to the volume of approximately one litre.

The reservoir for the brake fluid is located on the front right side of the cabin and the fluid level should be at the mark shown on the reservoir. If any part of the system is disconnected, the brake fluid should be completely drained and replaced.

Impact

In April 1998, the *Hands On* programme about electric three-wheelers, 'Safa Tempos', was screened on television for the first time. Evidence has shown that this programme influenced people and encouraged them to lobby the government to ban the diesel tempos.

The electric vehicle business started in 1996. Two years later, EVCO was one small fleet comprising ten vehicles. By 2001 the enterprise had grown to more than ten times that size. There are 10 EVCO depots and charging stations located throughout Kathmandu, providing more than 200 jobs.

Companies from India have taken an interest in the electric vehicles and some of them have visited Nepal to see how the vehicles are charged and how the electrical components are installed. Two of these companies have started to develop electric vehicles in India for themselves.

The people living in Kathmandu have benefited significantly from the use of electric vehicles. Air pollution has reduced and the air quality has improved.

Further information

Nepal Electric Vehicle Industry
PO Box 8975
EPC 5154
Lazimpat
Kathmandu
Nepal
Tel: +977 1 412 183
Fax: +977 1 415 055

Electrical Vehicle Company Limited
PO Box 9219
Bhat Bhateni
Kathmandu
Nepal
Tel/Fax: +977 1 420 670
E-mail: evco@mos.com.np

Self-service car rental

The city of La Rochelle, on France's Atlantic coast, has long been a pioneer in urban environmental protection and city-friendly public transit systems, but here as elsewhere there is increased vehicle congestion and urban pollution, affecting the quality of life for those living and working in the city. In view of this and its financial consequences for the city, the authorities decided to install a fleet of self-service electric cars.

By providing easy access to self-service rental of electric cars, a programme known as 'Liselec' offers a solution to the main issues faced by local authorities. Because no two cities are alike, the process includes an in-depth study of the city's demographic, sociological and urban features, and a joint analysis of existing transit possibilities and traffic flows. Commercial operations began in September 1999.

Guaranteed parking

There are 50 vehicles available round the clock at six recharging stations sited close to high-use locations, such as the railway station, the university, the cultural centre, the sports complex and the central shopping area. The cars have the advantage of being guaranteed a parking space at each recharging station because they are shared by a number of drivers, which optimizes the use of parking areas.

Subscribers access the cars with smart cards, which in the future may be valid for other kinds of transit systems. The programme has proven highly successful with the inhabitants of La Rochelle, who appreciate not only the system's ease of use, flexibility and reliability, but also the opportunity to enjoy driving an electric car.

Control centre

There is a control centre to supervise the rental system, which:

- sells contracts
- deals with customer relations
- monitors fleet dispatch
- dispatches cars to different locations
- carries out cleaning operations
- recharges vehicles.

Each recharging station is equipped with an electronic control unit and the cars record all their journeys.

Using the cars

Customers use a smart card to pay in advance for the use of one of the cars with no further administrative procedures. The transaction ends once the vehicle has been returned to one of the six centres.

1. Customers choose the monthly contract that suits their needs – either a fixed hourly fee with unlimited mileage, or a membership programme that bills for actual use. In each case, a wallet-sized smart card and a password are issued.

2. At the recharging stations customers will find a number of electric Peugeot 106s and Citroën Saxos. If the green light on the rear is lit, the car is charged and ready to go.

3. When the smart card is waved in front of the terminal, visible through the left rear window, the doors unlock automatically.

4. Once inside the car, the customer enters the password into the keypad to release the vehicle lock.

5. When the vehicle is returned to a recharging station, the password is re-entered into the keypad. To lock the door, the smart card is waved in front of the terminal.

Liselec: the self-service electric vehicle
Source: PeugeotCitroën SA

Specifications of an electric vehicle

The estimated lifetime of an electric car exceeds ten years. Unlike thermal engines, the motor of an electric vehicle does not require any periodic maintenance.

The range is 80 km. In a city a driver generally drives only about 30 km a day, so an electric car will need to be recharged every two or three days, preferably overnight.

For recharging, a simple 16A outlet is enough. An electric vehicle can be recharged at home, at night, or by using the numerous normal and rapid recharging points installed in the city. Drivers can recharge using their smart card.

Further information

E-mail: liselec@viagti.com
Websites: www.psa.fr/liselec/en_intro.html
www.psa.fr/en_index_enviro.html

Smart Cars

The motor car has been responsible for high levels of pollution since its invention at the beginning of the twentieth century. With growing concern about world pollution, the German company, Mercedes Benz, and the Swiss company, Swatch, decided to join forces and build a car that was as ecologically sound as possible. The Smart Car is designed by Micro Compact Car GmbH.

Lightweight city coupé

The Smart Car is a two-seater petrol city coupé which is lightweight and has been designed to take up less room in congested city centres. It has electronic power management that prevents the engine from emitting pollutants and it has an economy transmission mode to save fuel.

The Smart Car is assembled in modules, which makes design changes easier. For example, new power sources such as electric fuel cells could be fitted. It also guarantees that the car can be dismantled and recycled economically at the end of its useful life.

Specifications of the Smart Car

Fuel consumption: 4.8 litres per 100 kilometres
Length: 2.5 metres
Width: 1.51 metres
Height: 1.53 metres
Maximum speed: 135 kph
Acceleration: 0–60 kph in 7 seconds

The Smart Car has been designed to take up less room than other cars

New factory

The Smart Car is manufactured at the Smartville Energy Centre at Hambach in France. A completely new factory was designed with the quality of the environment in mind and the whole place is energy efficient. The factory was built on 30 hectares of industrial wasteland and now includes a nature reserve with rare plant species and landscaping for hundreds of trees. Building a car factory from scratch helps to make a green agenda possible but ecological conservation is still a complex process.

All the materials used in the construction of the Energy Centre were checked against a list of environmentally harmful materials. None of the buildings contains formaldehyde or CFCs. The panelling used for façades is made of TRESPA, which is a raw material produced mainly from the wood of a European tree species that recovers quickly.

Recycling water and wastes

Rainwater that runs from the roofs of the buildings is retained in reservoirs for use in tempering steel. All the other waste water that drains off the roads and car parks is fed into the oil separation plants, treated in storage basins and used for specific purposes.

A centrally located biological clarification plant purifies all the waste water from the factory's sanitary installations and industrial processes. The recycling of water is yet another way of preserving valuable resources. The plant works using new biomembranes that clean waste through a filtration system to strict European drinking water standards. After purification, the water is used in the gardens and as a coolant during the production process.

Energy saving and heat

The strict application of ecological principles is also clear from the way in which surplus materials such as excess powder are collected and reused. Energy saving is an important feature in the buildings, and insulation is used in the façades to muffle sound and conserve heat. Heat recovery systems are used throughout the factory. The heat generated by the injection moulding section and by air leaving the paint department is taken through a rotating heat recovery system. The use of waste heat to this extent is rare and eliminates the need for cooling towers as well as cutting costs and preserving resources.

The Smartville Energy Centre consists of a heating station and a block type thermal power station. Emissions are greatly reduced simply because natural gas is burned rather than conventional fuels and the efficiency of the system is boosted by reuse of waste heat.

Recycling materials

Recycled materials as well as raw materials that can be reproduced are used in the manufacturing process of the Smart Car. Risky materials are consistently avoided. For example, there is no use of poisonous metals, such as lead and cadmium, throughout the entire production process of the car. The result is that up to 95 per cent of the Smart Car is suitable for recycling.

Solvent-free production

The chassis of the new Smart Car is painted using the powder coating technique which is a solvent-free process. Apart from its ecological soundness, the process produces a higher-quality finish than conventional methods despite the fact that the coat is thinner. Another advantage is that there is no hazardous waste.

Environmental management systems

Environmental protection has been integrated into all the phases of development research for the production of the Smart Car and also in the planning of the factory and its systems.

An international environmental standard – ISO 14001 – aims to ensure that everything has as little impact on the environment as possible. Micro Compact Car GmbH has been awarded the Environmental Certificate ISO 14001 for the environmental protection measures it has used in the development of the Smart Car. Thus, the Smart Car is one of the first cars to be developed to the requirements of an environmental management system.

Further information

Daimler Chrysler
Tongwell
Milton Keynes MK15 8BA
United Kingdom
E-mail: queries@thesmart.co.uk
Website: www.thesmart.co.uk

Hydrogen power

One way to reduce carbon dioxide emissions could be by using hydrogen as a replacement for fossil fuels. Hydrogen is a 'clean' fuel because it produces virtually no pollution.

Hydrogen is the element with the lightest and simplest atom of all. It is a gas under normal conditions and it is the least dense of all substances. However, it has a very high energy content in relation to its weight, its energy density being three times that of petrol. It is also very abundant as it is one of the constituents of water, the other being oxygen. When it is burned in air, the only waste product is water.

There are many ways of producing hydrogen, but the most promising one from an environmental point of view is by using renewable energies, such as solar, wind or water power, to generate electricity which can then be used to break down water into hydrogen and oxygen. The oxygen can be safely released into the atmosphere and the hydrogen stored for use as a fuel.

Solar energy is seen as the most likely source of hydrogen production in areas where there are high levels of sunshine on a regular basis. The amount of energy falling on the earth from the sun every day is enormous. However, hydrogen has not yet been produced on a commercial scale.

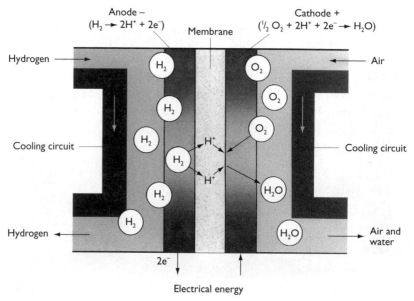

The fuel cell
© MAN

Hydrogen as a fuel

There are two main ways in which the use of hydrogen as a fuel is being explored. It is possible to burn hydrogen in an internal combustion engine in much the same way as petrol is burned. However, most research has concentrated on devices called fuel cells.

A fuel cell is a device which makes it possible to bring hydrogen atoms into contact with oxygen from the air in such a way that electricity is produced as the two combine to make water in vapour form. The amount of electricity generated in a single cell is quite small but many cells can be stacked together to make up a power unit for any particular application. One of the types of cell best suited to driving vehicles is the polymer electrolyte membrane cell, which operates at about 80°C and converts about 60 per cent of the energy contained in the gas.

The hydrogen bus

Several companies have come together in Germany to pool their expertise in the 'Hydrogen bus' development programme. A new fuel cell assembly provides a total of 120 kW of electrical energy to power the bus. The rear wheels of the bus are driven directly by two electric motors without the need for a gear change mechanism. The fuel cells produce direct current, so an inverter is necessary to convert it to alternating current. The inverter also provides a convenient method of controlling engine power.

The hydrogen fuel is stored in the form of compressed gas in

The hydrogen bus
© MAN

pressurized containers built into the bus roof. There are nine gas cylinders, together holding over 1500 litres of hydrogen fuel, enough to drive the bus for about 250 km.

Hydrogen-fuelled cars

A German car company is concentrating on another approach to the use of hydrogen fuel. Since it has many years' experience in the development and production of internal combustion engines, its development programme is based on its 750iL saloon car.

Modifications to the cylinder head allow the engine to burn either petrol or hydrogen and, on the present prototypes, it is possible to change from one fuel to the other at the flick of a switch. The company expects the final version of the car, which will be fuelled by hydrogen alone, to have a performance similar to the present petrol models but with the great advantage of having zero carbon dioxide emissions. The exhaust gases will consist of air and water vapour.

A number of hydrogen-fuelled cars are used in the district around Munich airport, providing a chauffeuring service between the airport and the city centre. An essential feature of the development programme is an automated refuelling station situated at the airport. The system is automated in order to provide maximum safety to the filling operation.

Further information

Ludwig-Bölkow-Systemtechnik GmbH
Reinhold Wurster
Daimlerstrasse 15
D-85521 Ottobrunn
Germany
Tel: +49 89 608 1100
Fax: +49 89 609 9731
E-mail: brunner@lbst.tnet.de
Information about hydrogen: www.HyWeb.de

Linde AG
Werksgruppe Technische Gase
Seitnerstraße 70
D-82049 Höllriegelskreuth
Germany
Tel: +49 89 74 46 1158/1223
Fax: +49 89 74 46 1230
E-mail: Dr._Thomas_Hagn@Linde-Gas.de
Website: www.linde.de/linde-gas

E-mail: info@brennstoffzellenbus.de
  info@fuelcellbus.com
Website: www.fuelcellbus.com

Hybrid power

Approximately 90 per cent of the environmental impact caused by a diesel-powered truck or bus occurs during the vehicle's working life. The production and scrapping of the vehicle together account for the remaining 10 per cent. The emission of carbon dioxide while the vehicle is in operation adds to the greenhouse effect. In densely populated areas and cities, it is important to reduce emissions of nitrogen oxides, carbon monoxide and hydrocarbons, which have a negative effect on both health and the environment.

As overcrowding and air pollution in our cities continue to escalate, Volvo, the Swedish motor manufacturer, is attempting to produce two vehicles that create 'the least possible environmental burden'. The Environmental Concept Bus and Truck have been developed as prototypes and are viable concepts, even if they are not commercially feasible at present.

Environmental Concept vehicles

Compared to today's vehicles, the Environmental Concept Bus (ECB) and the Environmental Concept Truck (ECT) produce negligible amounts of harmful exhaust emissions. These vehicles are powered by a hybrid system featuring a gas turbine with an integrated high-speed generator, batteries and electric motor. The gas turbine can run on virtually any liquid or gas fuel without necessitating major modifications. In this particular application, the engine has been modified for ethanol – a fuel that can be produced from renewable sources such as timber.

This is a solution that yields very small emissions of harmful exhaust gases. Emissions of nitrogen oxide are just one-tenth of those generated by a conventional type of diesel engine.

Hybrid power

The best way of combining sufficient operational range with a zero-emission facility is hybrid power where batteries are used for short stretches in particularly vulnerable environments such as city centres, with a combustion engine providing power at all other times.

The combustion engine in the Volvo series hybrid is a gas turbine which consists of a compressor, turbine, combustion chamber and heat exchanger which harnesses the heat in the exhaust gases and also acts as a silencer. These components combine to form a compact mobile power plant, the high-speed generator unit.

The vehicles offer two power alternatives – hybrid power and battery power. Under hybrid power, the high-speed generator (HSG) unit propels the vehicle via an electric motor fitted to the rear axle. When only a little power is required, the unit diverts all the surplus energy to the batteries for storage. When extra power is needed, the batteries release this energy to propel the vehicle at higher speed.

Construction

The frames of the bus and the truck are built entirely of extruded aluminium beams and these frames are covered in aluminium sheeting. The structure is both lightweight and strong. The production process requires large energy resources, but the recycling is easier and less energy-intensive than for steel.

The middle section of the bus is made of robust aluminium beams and there are sturdy steel roll-over cages at the front and rear which would prevent the roof from collapsing into the passenger compartment in the event of a roll-over accident.

A large proportion of the material used in the manufacture of the vehicles can be recycled and returned to the production process.

Suspension system

Active suspension in the vehicles contributes to improved safety and higher levels of comfort than found in many conventional vehicles. The advanced suspension system reacts instantly to compensate for any irregularities in the road surface and gives the vehicles excellent anti-roll stability. The floor height of the bus above the road surface is usually 320 mm but with the active suspension, it can be lowered to 170 mm to aid entry and exit.

Batteries

The vehicles are equipped with Nickel-Metal-Hydride batteries (NiMH) which currently offer the best performance in terms of environmental suitability and energy storage capacity. The NiMH batteries weigh half as much as lead-acid batteries and are far more environmentally friendly. They are maintenance-free with a long service life and they are designed to be recycled to about 95 per cent. In hybrid operations, the batteries are recharged very quickly by the HSG unit. They can also be recharged from the mains supply. The batteries are well protected in three separate units and the vehicle will continue functioning even if there is a fault in one of the units. The batteries provide 25 kilometres of zero-emission driving before they need to be recharged.

The electric motor not only propels the vehicle, it also functions as an electric brake retarder. This design allows braking energy to be used to charge the batteries instead of being lost as surplus heat. At the same time, the retarder provides gentle braking and reduces wear to brake linings.

Tyres

The tyres have been specially developed for these vehicles. Both the material and the production process are optimized for minimal environmental impact. The tyres' tread section is made without any environmentally hazardous oil additives.

The Michelin 'Super Single' rear tyres are 495 mm wide and provide 30 per cent lower rolling resistance than standard tyres. This means lower energy consumption and therefore lower exhaust emission levels. The front tyres offer about 15 per cent lower rolling resistance.

Mirrors

The rear view mirrors have been replaced by video cameras because they provide better all-round visibility. In addition to the side-mounted camera, there is a camera that monitors the rear of the truck and is activated as soon as reverse gear is engaged.

Lights and indicators

The bulbs in the indicators, position lights, tail lights and braking lights have been replaced by light-emitting diodes (LED). They consume just 20 per cent of the energy used by conventional bulbs and they last virtually forever. LEDs are far more durable under vibration and they light up more than 1000 times quicker. For brake lights, this faster activation time can give a following driver several more metres in which to react.

The headlamps and auxiliary lamps feature gas-discharge bulbs which offer

The Environmental Concept Truck

a more intense beam, long service life and low energy consumption. They are supplemented with UV lights, which offer twice the visibility of normal bulbs and work by reflecting off UV-sensitive substances found in textiles and minerals, etc.

Heating and air conditioning

The conventional source of heat in a vehicle is from the combustion engine's coolant. Cab heating in the concept truck and heating for the bus come from the coolant that carries away the surplus heat from the electronic systems, supplemented with electric heating. In order to reduce energy consumption and the risk of cold down-draughts, the cab of the ECT is well insulated and the door windows are double glazed.

The air conditioning system uses iso-butane refrigerant and it contains neither freons nor other chlorine-based chemical compounds which can impair the ozone layer.

Noise emissions

The noise level is low. The gas turbine produces a soft whining noise and when the battery power is activated, there is no noise at all.

Truck specifications

Wheelbase	5.3 metres
Overall length	10 metres
Height	3.3 metres
Width	2.3 metres
Front overhang	2.0 metres
Rear overhang	2.7 metres
Turning radius, outer	17 metres
Turning radius, inner	7 metres
Front axle load	5 tonnes
Rear axle load	10 tonnes
Gross weight	15 tonnes
Load height	1050 millimetres (+/− 150 mm)

The cab is a walk-through one and offers full standing height at 190 cm. It also has a floor height of just 60 cm above the ground. The driver sits at almost the same level as pedestrians, cyclists and cars, which increases the safety of the truck. Short and low, the ECT is built for dense city traffic.

Bus specifications

Wheelbase	8.4 metres
Overall length	10.7 metres
Exterior height	3.2 metres
Interior height	2.1 metres
Front overhang	1.25 metres
Rear overhang	1.05 metres
Turning radius, outer	10.7 metres
Turning radius, inner	5.1 metres
Width	2.55 metres
Gross weight	15 tonnes
Passengers, sitting	24–33
Passengers, standing	70–80

In a conventional bus, the power train usually has to be positioned below the floor, and this sets the limit on how low the bus can actually be built. This consideration does not need to be taken into account with the ECB because there is no gearbox or propeller shaft and all power transmission takes place electrically.

By placing the wheels way out in their respective corners, so they do not affect the interior passenger space in any way, and by fitting most of the power train components in the roof, the designers have created a very long wheelbase and a low and flat, uninterrupted floor. The ECB has been designed to accommodate the needs of disabled passengers and there is no difficulty in boarding the bus with either a wheelchair or a pram.

The short overall length of the ECB (1.3 metres shorter than most city buses) in combination with four-wheel steering and speed-dependent power steering means that it is extremely easy to manoeuvre. In the ECB, the driver sits directly above the front axle, midway between the two front wheels, which provides excellent vision.

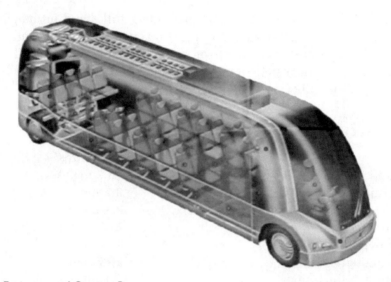

The Environmental Concept Bus

Further information

Michael Borg
Volvo Bus Corporation
Corporate Communications
Dept. 80900 VB1S
SE 405 08 Goteborg
Sweden
Tel: +46 31 66 64 52
Fax: +46 31 66 72 88

Bioengineering to prevent landslides

The Himalayas experience some of the highest rates of erosion in the world, made more acute by the humid subtropical temperature zones. Rapid rock weathering and heavy rainfall are predominant features of this area and are responsible for inducing landslides and erosion. Landslides block and cause damage to roads; farmland is swept away and houses are destroyed.

In Nepal, the road network has expanded rapidly since the 1950s and the design, construction and maintenance of roads must allow for the long and steep slopes which are subject to erosion, and very shallow slope failure. The slopes need to be stabilized to protect them against erosion, and so dramatically reduce landslides and the devastation these cause.

Bioengineering is the use of living vegetation for engineering purposes. It aims to protect and stabilize slopes by preventing erosion and shallow mass movement. It operates in the same way as civil engineering structures and carries out the same functions. Bioengineering offers an additional set of tools for the engineer, adding to the options available and increasing the scope of works. It is usually used in conjunction with civil engineering structures and non-living plant materials, rather than replacing them. There are limitations in using vegetation by itself – for example, deep-seated landslides, where the rupture plane is deeper than about 50 cm below the surface, cannot be stabilized.

Each slope needs individual assessment and consideration given to the angle, length, drainage and moisture levels. The skill in using vegetation in engineering is to combine it carefully with civil works to give the best results in terms of cost and effect. Bioengineering is relatively low in cost and the materials and skills are all available in rural areas. It takes some time for the materials to reach their maximum strength but, unlike civil engineering systems, they tend to become stronger over time.

Effects of vegetation on slopes

A cover of vegetation protects the soil against rain splash and erosion, and prevents the movement of soil particles down the slope under the action of gravity. Vegetation increases the soil infiltration capacity, helping to reduce the volume of runoff. Plants transpire considerable quantities of water, reducing soil moisture and increasing soil suction.

Stems and leaves cover the ground surface and absorb the impact of material moving down the slope, therefore protecting it. This network of surface fibres produces a tensile mat effect and restrains the underlying strata. During high-velocity flows, foliage is flattened and covers the soil surface, providing protection against erosive flows. The foliage also acts as a barrier to rainfall, creating a significant reduction in the kinetic energy of raindrops and thus diminishing their power of erosion.

Plant roots bind the soil, which increases the shear strength through a matrix of tensile fibres, resulting in increased resistance to deformation. The roots of plants penetrate deeply, giving anchorage into firm strata, bonding the soil mantle to stable subsoil or bedrock, and support to upslope material through buttressing and arching. These factors make a significant contribution to slope stability, but only once the roots are mature enough to reach deeply into the ground – which takes about five years.

Vegetation encourages other plants and animals to live on the slope and therefore bioengineering helps to improve the environment as well as providing useful products, such as firewood and fruit.

Characteristics of bioengineering plants

Vegetation can be selected and arranged on the slope to perform specific engineering functions. A mixture of plant types should be introduced so as to give a range of rooting depths and create an irregular structure. This tends to prevent continuous shear planes from developing in the upper soil layers, discouraging shearing from taking place. The plants on the site should be of different ages so that they do not all need to be replaced at the same time and to ensure that there will always be strong, healthy plants protecting the slopes.

Local species should be used rather than imported materials because native plants are more likely to be adapted to the growing conditions of the general environment and be resistant to local diseases.

An ideal bioengineering vegetation community has large trees which root deeply, giving the maximum anchorage effect. Shrubs with strong, woody roots which are shallower than the tree roots form an intermediate level and large clumping grasses, with a dense network of fibrous roots close to the soil surface, provide a thick surface cover to prevent erosion.

To increase the light penetrating through the canopy, the trees can either be 'pollarded' (as shown in the diagram) or 'coppiced'. Pollarding is where the main trunk is cut off about two or three metres above the ground and new, smaller shoots can grow. Coppicing is where the

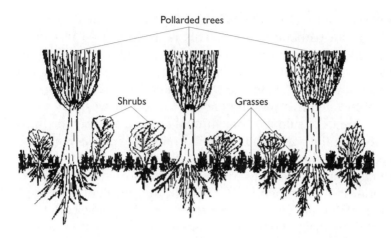

An ideal plant community for bioengineering should include different ages and rooting depths

trunk is cut off about 30 centimetres above the ground to allow new shoots to grow from the stump. Both treatments act as a form of thinning and the trees become lighter in weight and more flexible but retain their strong root systems. The plants growing on the ground improve because they are able to develop an extensive root network.

Functions of vegetation in slope stabilization

Plants are used for specific purposes in bioengineering and there are seven main functions for vegetation found on the steep, fragile and eroding slopes in Nepal.

- *Catching* material that is moving down the slope in the process of erosion, as a result of gravity or with the aid of water, is done using the stems of vegetation. It can be used on slopes where there is a risk of shallow failure or where there are other civil engineering structures, or in rehabilitation areas – for example, quarry or tipping sites – to catch materials or debris.
- *Armouring* the slope against surface erosion, such as runoff and rain splash, requires a continuous surface vegetation cover. Armouring can be used in areas where there is bare soil or where there is a risk of gullying on the slopes due to the surface material being weak or poorly protected against erosion. It can also be used as a slope component where other civil engineering structures are employed to armour the surface between inert structures.
- *Supporting* the slope from below, by propping from its base,

requires large, mature plants with deep, dense root systems, such as fully grown bamboo or trees. This is usually used in conjunction with other engineering works to improve support to the slope.
- *Reinforcing* the soil by increasing its shear strength. This depends on the strength and density of the plant roots. On slopes where the failure depth is less than about 50 centimetres, the most likely causes are debris flow or transitional slips, which can be limited by reinforcing the soil. Where general rehabilitation is required, reinforcement of the soil is often carried out.
- *Drainage* of the slope to avoid the slumping of saturated surface material is determined by the distribution of plants or the planted configuration on the slope; for example, using vertical or diagonal planting to direct water down the slope. The root systems of the plants can also be used as a drainage tool, by carrying water down into the soil as well as drawing it out by transpiration. The risk of shallow failure on slopes can be decreased by draining the soil, and vegetation can be used to drain excess runoff safely.
- *Limiting* the extent of slope failure can be achieved by using plant roots to hold the surface together.
- *Improvement* of the local environment, particularly the soil and microclimate, encourages better growth of other vegetation, either naturally or through management.

Bioengineering techniques for roadside slopes

Planted grass lines: Grass sprigs are planted in lines on the slope. The lines can be planted on the contour (horizontal), which will trap materials moving down the slope; or downslope (vertical), which maximizes surface drainage while protecting against erosion; or on the diagonal, which combines the benefits of horizontal and vertical planting.

Grass seed: Grass seed is spread over the ground surface to give complete but random surface armouring. It is often covered in mulch (the stems and leaves of unwanted plants are cut up and placed around seedlings to keep the soil cool and moist) to aid its development but it still takes considerable time for sizeable plants to grow.

Turfing: A surface is covered with turf, which gives complete and instant surface armouring, although this is a relatively expensive method to use.

Shrub and tree planting: Seedlings of shrubs and trees are planted at intervals throughout a site. This is usually feasible only on fairly shallow slopes. They grow to reinforce and anchor the slope but it will be about five years before they contribute significantly to slope strengthening. Care (and perhaps protection against grazing) will be required in the first three years of growth.

Shrub and tree seeding: The seeds of shrubs and trees are inserted into cracks on steep, rocky slopes and grow to reinforce and anchor the slope. These plants take about five years before they make a significant contribution to slope strengthening, and during this period need protection.

Large bamboo planting: Large clumping bamboos are planted, usually close to the base of the slope. This establishes a very strong line of plants which provide excellent reinforcement, trapping and support. Bamboos take about five years to contribute significantly to slope strengthening and require protection in the early years.

Brush layering: The lower ends of live woody cuttings (usually hardwood) are laid in shallow trenches across a slope, usually following the contour, and the aerial part is left sticking out above the ground. These cuttings form a strong barrier to prevent the development of rills and to trap material moving down the slope. Excess debris can roll over brush layering with minimal damage, which helps it to survive long enough to take root and grow into strong shrubs. If, the lines are angled correctly, brush layering can serve a drainage function.

Palisades: Live woody cuttings are planted in horizontal lines across a slope, usually following the contour, and make fences consisting of closely spaced upright cuttings. They provide a strong and low-cost barrier to slow the development of rills, trap material moving downwards and reinforce the soil with minimum disturbance to the slope. Palisades can be used for drainage if positioned at the right angle.

Live check dams: A variety of woody cuttings are used to build a live check dam, which can be an effective low-cost structure to reduce erosion in smaller gullies through armouring and reinforcement. They also trap materials moving down the slope and will act as drainage if positioned at the right angle.

Fascine constructions: Bundles of live branches are laid in trenches just below the surface, usually following the contour. This is a very strong and low-cost barrier to trap material and reinforce the slope once it has grown through the interlocked root system. It can also be used for drainage if it is angled at the right position.

Choosing the right bioengineering technique

Slope angle	Slope length	Material drainage	Site moisture	Technique(s)
> 45°	> 15 metres	Good	Damp	**Diagonal grass lines**
			Dry	**Contour grass lines**
		Poor	Damp	1 **Downslope grass lines and vegetated stone pitched rills** or 2 Chevron grass lines and vegetated stone pitched rills
			Dry	**Diagonal grass lines**
	< 15 metres	Good	Any	1 **Diagonal grass lines** or 2 Jute netting and randomly planted grass
		Poor	Damp	1 **Downslope grass lines** or 2 Diagonal grass lines
			Dry	1 **Jute netting and randomly planted grass** or 2 Contour grass lines or 3 Diagonal grass lines
30°–45°	> 15 metres	Good	Any	1 **Horizontal bolster cylinders and shrub/tree planting** or 2 Downslope grass lines and vegetated stone pitched rills or 3 Site grass seeding, mulch and wide-mesh jute netting
		Poor	Any	1 **Herringbone bolster cylinders and shrub/tree planting** or 2 Another drainage system and shrub/tree planting
	< 15 metres	Good	Any	1 **Brush layers of woody cuttings** or 2 Contour grass lines or 3 Contour fascines or 4 Palisades of woody cuttings or 5 Site grass seeding, mulch and wide-mesh jute netting
		Poor	Any	1 **Diagonal grass lines** or 2 Diagonal brush layers or 3 Herringbone fascines and shrub/tree planting or 4 Herringbone bolster cylinders and shrub/tree planting or 5 Another drainage system and shrub/tree planting
< 30°	Any	Good	Any	1 **Site seeding of grass and shrub/tree planting** or 2 Shrub/tree planting
		Poor	Any	1 **Diagonal lines of grass and shrubs/trees** or 2 Shrub/tree planting
	< 15 metres	Any	Any	**Turfing and shrub/tree planting**
	Base of any slope			1 **Large bamboo planting** or 2 Large tree planting
Special conditions				
Any*	Any*	Any*	Any*	**Site seeding of shrubs/small trees**†
> 30°	Any	Any rocky material‡		**Site seeding of shrubs/small trees**
Loose sand		Good	Any	**Jute netting and randomly planted grass**
Heavy clays		Poor	Any	**Diagonal lines of grass and shrubs/trees**
Gullies ≤ 45°	Any gully			1 **Large bamboo planting** or 2 Live check dams or 3 Vegetated stone pitching

Notes: * Possible overlap with parameters described in the rows above.
  † May be required in combination with other techniques listed on the rows above.
  ‡ Material into which rooted plants cannot be planted, but seeds can be inserted in holes made with a steel bar.
  Techniques in **bold type** are preferred.
  Chevron pattern: <<<<<   Herringbone pattern:←←←←←

Managing bioengineering communities

The vegetation needs to be managed, otherwise it may not carry out the functions required for engineering and might even have a negative impact on the protection and stabilization of the slopes. Unmanaged vegetation tends to produce a dense canopy of trees, which reduces the amount of sunlight reaching the grasses and shrubs. The trees may have deep roots, but with relatively little vegetation able to survive beneath them they are not able to stop erosion on the surface. They may also grow too big and start to destabilize the slopes on which they are growing by toppling in strong winds.

The aim is to establish a vegetation community that does not need too much intervention from outside to maintain it: for example, plant species that regenerate naturally; species that do not grow too fast or too tall (thereby reducing the need for frequent cutting and removal); and species that live longer.

During the initial implementation, the vegetation on the slope should be trimmed back and the loose soil chipped off and removed. The plants are prepared ready for use and a line is plotted for them to take. After planting, the slopes must be kept in good order, with regular maintenance to remove undergrowth and weeds. The plant nursery must be kept well stocked with the right species for the next planting season.

The information provided is a summary explanation of bioengineering, which is extremely site specific. The decision to implement one of the alternatives documented requires careful consideration of a wide range of situation-specific parameters, many of which are not addressed here.

Further information

John Howell
Living Resources Limited
Durston's Field
Heath House
Wedmore
BS28 4UJ
United Kingdom
Tel: +44 (0) 1934 713656
Fax: +44 (0) 1934 713658
Email: Living_/Resources@compuserve.com

Geo-Environmental Unit
Department of Roads
Babar Mahal
Kathmandu
Nepal
Tel/Fax: +977 1 231981
Email: geu@meu.gov.np

Chapter Five
Building a safe environment

Shelter is one of the basic universal requirements, and the poorest members of society need to find new accommodation more often than most, as they are the most likely to be afflicted by drought, fire or floods. They are also typically in the least desirable locations, such as next to motorways or railway lines.

In Kenya, Maasai housing has been modified to make it more durable and more comfortable, and by using a concrete skin to make the roof suitable for collecting rainwater for household use. In South Africa, low-cost housing schemes – created to rehouse those living in informal settlements – were considerably improved when the owners received training to enable them to help with building and maintenance, thereby reducing costs, and were involved in decision making.

Fly-ash, a waste product from the power industry, has been mixed with lime and gypsum to create strong blocks for building in a cyclone-prone area of India, and this project also incorporated training of local engineers and artisans in the building process, which identified cyclone-resistant designs based on traditional forms. Lime cement, another traditional material which had fallen into disfavour only to be revived in recent years, has been produced in Malawi in a small kiln.

Earth is a traditional building material which allows the local population to take charge of their built environment, and considerable work has been done to show how attractive such buildings can be. Some natural materials, such as bamboo, are regarded as 'poor' and have to be disguised to be acceptable to some communities, as in the example described here from Colombia, despite the fact that bamboo housing withstands earthquakes better than brick or concrete structures, and so is particularly appropriate to earthquake-prone areas.

The technique of strapping roofs of Jamaican buildings against hurricanes is described, as is low-cost brick production in Zimbabwe, made possible by using boiler ash, a waste product from power stations, for firing the bricks.

Finally, there are noise walls, designed in the Netherlands to protect housing constructed close to motorways or railway lines from traffic noise that would otherwise affect inhabitants' quality of life. Such houses are thus made more habitable.

Improved traditional housing

The Maasai have traditionally been pastoralists leading a nomadic life, moving from place to place with their cattle in search of better grazing land and good sources of water. In recent decades, however, the Maasai have been forced to lead more settled lives. It is impossible for them to keep on moving because of land subdivision and their traditional nomadic lifestyle is changing to a more sedentary one.

The Maasai women have always been responsible for constructing, maintaining and managing their homes, called *enkangs*. Until now, their settlements had always been temporary and were used mainly for sleeping and cooking in. The *enkangs* were built around the central cattle kraal and, along with the peripheral fences, formed the security barriers for cattle. When the time came to move from one grazing area to another, the settlement structures were left to decay.

Enkangs are made of poles, twigs, and grass, and plastered with cow dung and mud. They are characterized by low, leaking roofs; damp, smoky and dark rooms; cramped space; animal odour; lack of security; weak, termite-infested foundation posts and lack of durability. Ventilation is channelled through a narrow opening which serves as the entrance, and some *enkangs* have one or two 'windows' – holes in the wall of no more than 20 × 20 cm. The roof and the walls frequently crack and peel, requiring constant maintenance.

Enkangs are uncomfortable, lack privacy, and require constant time and effort to maintain. They are susceptible to fire, pests and harsh weather, and pose health risks – particularly eye and respiratory ailments.

Appropriate technologies

More durable and permanent houses are now required, but any changes to the houses must respect the cultural traditions of the Maasai. The women plan and redesign their own homes, using the same internal designs that exist in the *enkangs*, which provide the physical setting for family rituals. Any adaptations to the technology respect and maximize the women's indigenous building skills, such as plastering, and remain within the reach of local resources.

The house improvements aim to:

- reduce the need for time-consuming maintenance tasks by preventing roof leakage

BUILDING A SAFE ENVIRONMENT CHAPTER FIVE **157**

- reduce water collection time and improve water quality through rainwater collection for domestic use
- improve the living environment by improving natural lighting in the home
- reduce smoke levels and improve health through improved ventilation within the home.

Before and after improved Maasai housing

Following tradition, the Maasai women work together when they are building a new home. As in the past, the main frame of the house is built from wood and the walls are plastered using cow dung.

Concrete skin roof construction

A thin layer of cement mortar is used to form a concrete skin roof over the existing wattle and mud with dung roof. It is laid on a base constructed from twigs, grass and compacted soil. A polythene sheet is put between the base and the concrete layer to give protection against leakages and act as waterproofing. Chicken wire is used to further reinforce the mortar.

If the skin roof is made for a new house, only twigs should be used because grass takes time to settle. Using grass on a new roof could cause sagging and lead to water traps forming, which may produce leaks.

The roof incorporates guttering which channels rainwater into a small ferrocement water collection tank. During the rainy season, this is an efficient way of collecting water. It also relieves the women of the repetitive work of plastering the roof with cow dung after every rainy season.

Raw materials

Cement mortar: Sand and cement are mixed in the ratio of 3:1. The cement should be a fine powder and stored in a dry place, above the ground. The sand should be well graded and come from a pure source,

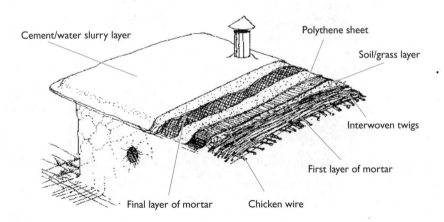

Structure of the concrete skin roof

with medium-sized coarse grains to enable the cement to bind with it properly. If it contains big stones, it should be sieved to remove them.

Water: The water must be clean and not from a stagnant or polluted source.

Wire: Different sizes of chicken wire can be used for the skin roof. The smaller the hole sizes, the more expensive the wire tends to be. To avoid rusting, the wire should be kept in a dry place until it is ready to be used.

Plastic sheeting: The sheeting acts as waterproofing and holds back any cement mortar that may seep through gaps in the soil layer.

Waterproof cement: Mixed with ordinary cement in the ratio of 1:1 and used as a finishing layer to make the skin roof waterproof.

Tools

Very simple tools are needed for the construction of the roof. They include sieves for removing large particles of sand; spades for mixing the cement and sand; pliers for cutting and folding wire; batch boxes or karais for measuring quantities of raw materials; and trowels for laying and smoothing the mortar.

Benefits of the improved housing

The new *enkangs* are more durable and are capable of withstanding extreme weather conditions. The design, with wider entrances and

The improved Maasai house with skin roof, integral guttering and ferrocement water jar

increased roof height, allows for a flexible internal layout. The natural lighting and ventilation are much improved and the fire risk has been reduced. The risk of domestic accidents has been minimized by the increase in lighting and extra space. More activities can be comfortably carried out inside the house. Improved ventilation and wall heights have lessened the rate of eye and chest infections (from smoke) and backaches (from constant bending). Air circulates more easily and there is less heat inside during cooking.

These *enkangs* have reduced the amount of time women spend on repair and maintenance of their homes, and the time spent on water collection from faraway places has also been significantly reduced. This has allowed focus on other important concerns such as childcare, health, nutrition and kitchen gardens. Women have been able to use their time effectively in income-generating activities such as bead making.

Further information
ITDG Kenya
Chiromo Access Road, off Riverside Drive
PO Box 39493
Nairobi
Kenya
Tel: +254 2 442108/446243/443710
Fax: +254 2 445166
E-mail: itdg@tt.gn.apc.org

Low-cost concrete housing

One of South Africa's most pressing problems is the provision of suitable shelter for the huge numbers of people living in shacks in the sprawling settlements. These shacks are constructed from anything that can be acquired, such as scrap timber or old roofing sheets and, while they may give some protection from rain and the heat of the sun, they do not provide adequate housing. The poor materials and the makeshift way in which they are constructed can also lead to a risk of fire.

The construction of low-cost houses in the township of Khayelitsha is based on locally made building components and has been made possible by the development of equipment which is itself low in cost.

The housing project for townships in South Africa

Work on the project started in 1994 as a cooperative venture between the International Development Group (IDG) of Birmingham School of Architecture and the Margarette Pierson Research Trust (MPRT), with the support of the United Kingdom's Department for International Development (DFID).

The project included a study, at an early stage, of various low-cost housing schemes in South Africa, built by contractors using standard building materials. The study found that most were built to very low standards and were, in some respects, worse than the shacks they were intended to replace. The study concluded that many of these housing schemes produced houses that:

- are neither structurally sound nor suitable for living
- have high maintenance requirements due to poor construction
- have no control by, or contribution from, the owners.

The study indicated that the involvement of the owner as unskilled labour could have saved more than 12 per cent of the cost, which could have been well spent on improved standards of construction.

The project avoids these old failings by using construction materials and techniques that allow good quality houses to be built. By actively encouraging the involvement of the community in all aspects of the process – decision making, materials production and construction – the project ensures that family members learn new skills. This allows them to contribute labour to the building of the home and also makes it possible for them to take control of the maintenance of the home.

The first part of the work concentrated on the development and testing of pedal-operated equipment and a moulding system for the production of a range of building products. The equipment was used to make concrete blocks, roof arches, floor slabs, tiles and roof sheets.

When this production system had been developed, it was possible to devise a low-cost house building system that was intended to:

- facilitate self-help housing
- meet the needs and expectations of the people
- be simple and require minimal skilled labour
- be affordable to low-income families.

The complete system was then brought to the attention of the people of Khayelitsha township by means of organized training workshops, which covered the production processes and quality testing methods for the building components, as well as the construction methods. The trainees were involved in the construction of a model low-cost house in the township in March 1995.

The building components

The components used to build a range of different houses are all the same and can be made by families after suitable training, but the project team recommended that they should be made in a community production centre so that proper quality control could be ensured. Such a centre would also help to minimize production costs.

The *building blocks* for wall construction are 390 mm × 200 mm × 200 mm and weigh about 21 kg each. They are cast in a metal mould and vibrated for one minute on the pedal-driven table, before being left to cure for at least three weeks. As with all concrete curing, this stage must be done in a damp environment to allow the concrete to develop its full strength.

Concrete *paving slabs* measuring 400 mm × 400 mm × 25 mm are cast in plastic moulds and vibrated for 30 seconds. These must be kept in their moulds for 24 hours before being put on to a flat surface for curing.

Roof construction

The roof construction is based on two components:

The roof *'T' beam* is 4 m long and 160 mm high and has a weight of about 130 kg. It is cast in a metal mould and is reinforced with three metal bars 14 mm in diameter. It is left in the mould for at least 24 hours before being turned out on to a flat floor for curing.

The *roof arches* are 700 mm × 500 mm × 18 mm and each weighs about 17 kg. They are made in a metal frame on the vibrating table and then slid on to the arched moulds where they remain for about 24 hours. When they are strong enough to be taken from the moulds, they are stacked in a tank of water to cure for a week. They then require a further two weeks' curing in a shaded area and must be wetted twice each day. This careful curing ensures that the concrete develops its full strength – which it will need in the construction process as the arches have to support the upper concrete layer until it has set fully.

House construction

The house walls are built up in the normal way using the concrete blocks and when they are complete a wooden mould (known as shuttering) is fixed all round the top so that concrete can be poured on to the blocks to form the 'ring beam'. When this is set, the 'T' beams are placed across the house at 80 cm intervals so that they rest on the ring beam. The 'T' beams are carefully lined up so that they are parallel with each other and then the arches can be placed in position. The structure is completed by casting wet, lightweight concrete directly on to the arches. The roof is then completed with a waterproof outer coating.

A number of different types of house can be constructed with these components and most have been designed so that they can be extended later when more space is needed or when the owners can afford to improve the family home.

Finance

As part of the project's design, another study focused on the affordability of home ownership. It looked at the different income levels in the townships and calculated the loans that they could sustain over a five-year repayment period. This information helped to guide the designs of the houses themselves. The result was a range of home designs suitable for most incomes within the community. Each type of house can be constructed in several phases so that it can grow with the needs and income of the family.

Further information

Dr Mohsen Aboutorabi
International Development Group
Birmingham School of Architecture
University of Central England
Perry Barr
Birmingham B42 2SU
United Kingdom
Tel: +44 (0) 121 331 5134
Fax: +44 (0) 121 331 5131
E-mail: mohsen.aboutorabi@uce.ac.uk

Earthen architecture

In most countries of the world it is possible to mould earth with a variety of tools to construct buildings. The range of technical, constructional and architectural possibilities is extremely wide, and this has enabled the construction of modest shelters, village houses, urban blocks and religious structures, as well as palaces and entire cities.

In countries with no industrialized means, in a wide range of latitudes throughout the world, earth remains the main building material. Processed materials are costly both in foreign currencies and in imported fuels. Communities remain dependent on the use of locally available solutions, materials and knowledge. These materials and techniques are generally very well used and can ensure true architectural quality which makes the most of the human and material resources available.

Building techniques using unbaked earth

Earth is a ready building material that requires little processing. The techniques mainly associated with processes using moulds, shuttering and direct shaping are 'adobe', 'rammed earth', 'straw clay', 'wattle and daub', 'cob' and 'compressed blocks'.

Wattle and daub: A supporting frame, usually wooden, is filled with a daubed lattice or netting woven from vegetable matter. A very clayey earth is used, mixed with straw or other vegetable fibre to prevent shrinkage upon drying.

Straw clay: The soil used is very clayey and is dispersed in water to form a greasy slip which is then added to the straw. The earth binds the straw together. Straw clay can be easily adapted to the prefabrication of various building components, such as bricks, insulating panels and flooring blocks.

Cob: Balls of earth are stacked on top of one another and lightly tapped with hands, feet or tools to form monolithic walls. The earth is reinforced by the addition of fibres, usually straw from various types of cereal or other kinds of vegetable fibre, such as grass and twigs.

Adobe: (sun-baked earth brick) The bricks are made using a thick malleable earth, often with straw added. Traditionally, adobes were shaped by hand, in wood or metal moulds, but nowadays the use of machines is widespread.

Rammed earth: The earth is compacted in a framework. Traditionally, wooden forms and rammers are used, but in many countries steel forms and pneumatic rammers are common. It is possible to build monolithic walls with the compacted earth.

Compressed earth blocks: The process of compressing earth blocks has been mechanized, and manual or hydraulic presses or completely integrated plants can be used. Products range from accurately solid shape, cellular and hollow bricks, to flooring and paving elements.

Low-cost compressed earth block housing
© T. Joffroy

Adobe, rammed earth and compressed earth blocks are the most widespread earth construction techniques. They have reached extremely high scientific and technological levels, and permit the construction of a wide variety of components and systems – for example, foundations, floors, pitched and flat roofs, arches, vaults, domes, tiles, chimneys, canals, roads, dams and bridges.

Clay as a binder

Clays, in their unfired state, are the main binders of earth. They are deeply embedded in traditional building cultures in many parts of the world. Although they have the limitation that they soften when wet, they are the cheapest binders, with very low energy consumption. It is estimated that over a third of the world's population lives in houses of earthen construction.

Buildings of unstabilized earth face the risk of erosion unless special design measures are taken to reduce exposure to rain and moisture. For durability, earth should be used only where it is not prone to damp. Optimum designs will depend a lot on the environment, such as the natural drainage and water table; on the climate – for example, rainfall (quantity and intensity) and winds during rains; and on the maintenance practices of the users.

Stabilizers and other additives or methods such as good compaction and grain size optimization can reduce swelling, shrinkage and cracking, increasing strength and water resistance.

When clay is mixed with water it becomes malleable, plastic or liquid, allowing it to be shaped. When drying, clay sets and recovers its cohesive properties, and so can bind the soil together.

Most soils consist of clay together with proportions of silt, sand and

Pavilion of the Royal Commission of Jubail and Yanbu at Janadriyah, Saudi Arabia
© T. Joffroy

gravel. The larger particles give structure to a soil, while the clay holds it together and to a great extent provides the cohesion.

To obtain a good building material which is strong and easy to use, the proportion of clay in a soil should be about 15 per cent on average. The sand should be 40 to 80 per cent, the gravel up to 40 per cent and the silt 10 to 25 per cent. If the clay content in a soil is too high, minerals such as sand and gravels, or fibres such as straw or hair, can be added.

Generally, a fairly wet mix with higher proportions of clay is used in mouldings and spreading applications, while a mix with less clay is best suited to compaction in a moist or damp state.

Environmental advantages

- Unbaked earth does not contribute to the deforestation that follows the use of organic resources for firing baked earth materials.
- It does not consume any non-renewable energy – for example, oil and gas – for the processing and production of materials or for their application, as does the production of cements, lime and other conventional binding materials.
- By exploiting strata on construction sites, it allows a considerable saving in energy for the transportation of materials.
- It does not contribute to the degradation of the landscape as does the extraction of minerals and ores which hollows out hillsides and opencast sites. A great deal of the earth excavated in the course of large public facilities work – for example, roads – can be recycled and used in building.
- It does not require the excavation of aggregates, such as gravel and sand from quarries or from water courses, in insular sites or lagoons, putting into peril the ecological balance of these natural environments.
- It uses very little water, essential for the life of the people.
- It produces no industrial or chemical waste and, moreover, has the additional advantage of being almost entirely recyclable.
- Unbaked earth is not only non-polluting in its use, it also guarantees the absence of harmful effects in the context of daily life such as the absence of gaseous emissions or other toxic chemical components, radioactive emissions, and so on.
- The surface texture, colour, form and luminosity of unbaked earth make it an attractive material for buildings without ruining the natural environment.

Economic advantages

- Unbaked earth is often comparable in cost with – or indeed, more economical than – competing technologies. It requires no major financial transport costs because of its generally light production infrastructure.
- Unbaked earth requires only simple production and application tools (moulds, presses, light shuttering and masonry tools, etc.) which are accessible to a wide population of masons and self-help builders.
- Unbaked earth follows on in the heritage of the traditional architecture of numerous countries using local materials. It allows local populations to take charge of the production of their built environment and so control their living environment.

Further information

CRATerre-EAG
Maison Levrat
Parc Fallavier
Rue de la Buthiere
BP53
F-38092 Villefontaine Cedex
France
Tel: +33 4 74 95 43 91
Fax: +33 4 74 95 64 21
E-mail: craterre@club-internet.fr
Website: www.craterre.archi.fr

Cost-effective school buildings

Currently, in two-thirds of the states in India, literacy levels are below 20 per cent. With the rapidly expanding population, the need for schooling and primary education is greater than ever before. Until recently, most schools were built by engineers whose methods, materials and techniques ignored the skills potential and resources of local artisans.

In 1994, the Department for International Development (DFID) and the British Council began an imaginative training programme to help engineers, local artisans and communities to find innovative solutions to meet the demand for schools and reduce the cost of the infrastructure. It was required that only materials that minimized the release of pollutants to the atmosphere during the production process should be recommended for the construction of the schools, and that they and the fuel needed to produce them should have little or no adverse effect upon the building environment.

Foundations

Primary schools in Andhra Pradesh are generally single-storey structures. The type of foundation used is dependent on the soil conditions. Foundations distribute the entire building load on soil in such a manner that no detrimental settlements take place.

Wall foundations
For the construction of wall foundations, a trench measuring 830 mm wide and 75 mm deep is excavated and 1:5:10 lean concrete is levelled over it. On this base, 680 mm wide and 150 mm deep 1:3:6 plain cement concrete is laid. On this, 380 mm wide locally available coarse rubble stone blocks are laid with 1:6 cement mortar.

Arch foundations
In an arch foundation, the walls are supported on brick or stone masonry arches springing from a series of square cement concrete bases. This method helps to reduce the use of materials such as cement, coarse rubble stone, sand, etc. It is also a labour-intensive system and requires a high degree of supervision. Arch foundation is best to use if the condition of the soil is good and the building is low rise.

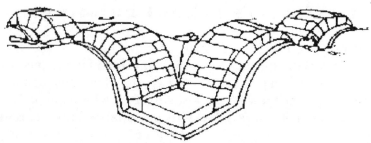

Arch foundations

Walls

Rat-trap bonded brick masonry
In rat-trap bonded brick masonry the bricks are placed on their edges in 1:6 cement mortar and, after the first layer of bricks has been laid, a gap is left between the bricks in the remaining courses. This means that, compared with a 230 mm thick solid brick wall, the amount of bricks required to build the wall is reduced by 25 per cent and, consequently, the amount of cement mortar needed is also reduced. The gap in the bricks helps to create thermal insulation, although it is not a good sound insulator.

Cement-stabilized mud block masonry
Cement-stabilized mud block (CSMB) is made by compacting a mixture of soil and cement in a blockmaking machine. The machine is usually operated manually and requires no fuel for production. The blocks are used in masonry in a similar manner to solid conventional brick work, but the mortar used is stabilized soil. The local soil needs to have the following range of particle distribution for it to be an effective component in the creation of CSMBs – gravel: 0–10%; sand: 40–70%; silt: 15–25%; and clay: 8–25%.

This type of masonry is effective where other walling materials are expensive, or of poor strength, in areas with medium to low rainfall.

Roofs

Micro-concrete roofing tiles
Micro-concrete roofing (MCR) tiles are made from locally available raw materials, such as sand, fine gravel, and cement using small stone chips. These tiles produce a high-quality, low-cost roof which is cheaper, cooler and quieter than corrugated iron sheets or asbestos. Roofing with MCR tiles is similar to conventional Mangalore tiled

roofing construction. It needs under-structures, such as rafters and purlins, which can be wooden, ferro-cement or steel, to support the tiles.

The tiles are produced using a self-contained vibrating device which can operate from mains electricity through a transformer, or direct from car batteries or solar power, or it can be hand powered. A plastic sheet is clamped on to the vibrating table and a measured scoop of mortar mix is placed into the frame. The vibrator is switched on and the mortar is trowelled flat. The frame is removed and the wet screed is carefully drawn on to a plastic mould. It is left to harden overnight and then placed in a solar curing area for three or four days, or is cured underwater for up to two weeks.

For MCR tile production there must be access to a vibrating table, a number of moulds, cement stone chips, sand, water and electricity. The tiles can be used in all situations where Mangalore tiled or asbestos roof is currently used. It develops local skills and entrepreneurship, and generates income. However, it is not as strong as a reinforced concrete slab and thermal insulation is low.

Concrete filler slabs

If the local Mangalore tile is used as a filler to reinforce the concrete slabs (see diagram), only a few tiles are required to build the roof. Mangalore tiles are placed between steel ribs and concrete is poured into the gap to make a filler slab. Less steel and cement is required in this structure and it is also a good heat insulator.

Corbelled brick pyramid

Some roofs have replaced tiles and slabs with bricks. A brick roof is an ancient traditional Indian construction technique and it uses less energy in construction than a conventional concrete roof.

A brick placed on top of another one, with an overhang of less than

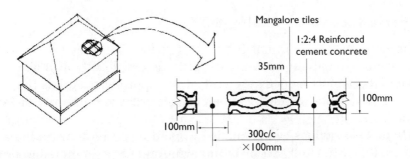

Filler slab with Mangalore tiles

Corbelled brick pyramid roof

half the length of the brick, will not topple and is called a corbelled brick. A corbelled pyramid is developed by creating layers of corbelled bricks.

Corbelled brick arch

When a brick arch is constructed by corbelling each brick from an existing arch, then it is called a corbelled brick arch. If a series of corbelled brick arches are constructed from the two opposite corners of a room so that they meet at the diagonal of the room, then it is called a corbelled brick arch roof.

Corbelled brick pyramid, dome or brick arch roofs are suitable for hot, dry or humid climates and cyclone-prone areas. Roofing of this kind uses comparatively small amounts of cement and steel and creates a large volume of space.

Cost-effective technologies

The new methods can cut the cost of school buildings by up to 30 per cent through employing local artisans and locally available raw materials, which reduces the cost of transporting materials. Repairs and maintenance also tend to be cheaper because of the use of local resources.

More cost savings can be made by avoiding building windows and doors. It is estimated that installing windows can cost up to ten times more than the installation of a wall. Instead, grills for light and air movement can be created in the brickwork. This is known as *jali* in India.

Note: The techniques and materials described here are only a small representation of the options and technologies available for cost-effective building construction.

Further information

Department for International Development
British Development Cooperation Office
50 M, Shanthi Path
Chanakyapuri
New Delhi – 110 021
India

British Council Division
British High Commission
17 Kasturba Ghandi Marg
New Delhi – 110 001
India

Hurricane-resistant roofing

Homeowners in the lower income bracket in Jamaica are most at risk during a storm or hurricane because many live in wooden houses that are generally in a state of disrepair and are not structurally sound. The houses are not attached to the foundations and they do not have roof ties or hurricane straps which would make a building more resistant to strong winds.

When Hurricane Gilbert hit Jamaica, it caused millions of dollars' worth of damage. For those living in informal settlements and poor housing the devastation was catastrophic.

The Construction Resource and Development Centre

A retrofit programme was set out by the Construction Resource and Development Centre (CRDC) and work was carried out with local roof builders who learned the principles of hurricane-resistant roofing. Safer roofs are still being installed.

Specifications for a roof

Roofs are held together by five major connections:

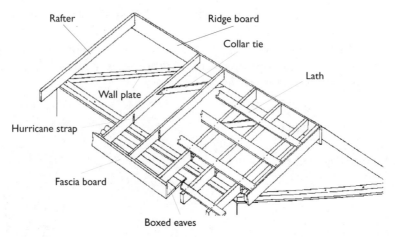

Parts of a roof

- the wall plate and the walls
- the rafters and the wall plate
- the ridge board and the rafters
- the laths and the rafters
- the roofing sheets and the laths.

If these connections are not installed properly, it will result in partial damage or complete loss of the roof during a hurricane.

Wall plate and walls

The wall plate is the first connection between the walls and the rest of the roof and it forms a frame on which the other sections of the roof sit. It must therefore be securely fastened to the rest of the building or the entire roof will lift in a hurricane.

For new roofs, wall plates should be held down to the block work using long bolts (10 mm × 200 mm) with washers, approximately 1.2 m apart. The bolts should be placed at least 120 mm into the belt beam, leaving at least 50 mm of the bolt remaining above the belt beam to fit the wall plate. Bent reinforcing steel should not be used because it can be straightened during a hurricane and when this happens, the wall plate will lift from the top of the building. Continuous lengths of timber should be used to make the wall plates and each piece should be fastened with at least two bolts or fixings.

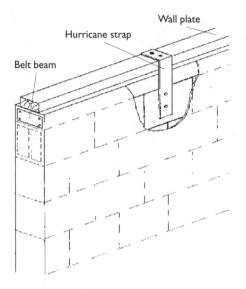

Strapping down the wall plate

To strengthen an existing roof, rawl bolts should be drilled into the belt beam and placed 1.2 m apart. Metal straps made from steel sheeting (25 mm × 5 mm) can be placed over the wall plate and fastened to the blockwork.

Rafters and wall plate

Rafters are usually made from 50 mm × 100 mm deep timber that runs from the eaves to the ridge board. The rafter is connected to the wall plate and the ridge. Twisted hurricane straps should be installed where the rafters join the wall plate. They should be nailed or screwed to both the wall plate and the rafter, thereby preventing the rafters lifting off the wall plate. Old zinc sheeting cut into strips can be used instead of straps. The sheeting should be cut into 25 mm wide strips and nailed over the rafter and into the wall plate.

By raising the roof to increase the slope, the pressure on the rafters is reduced and if the overhang is kept to less than 450 mm, the roof is more likely to remain intact.

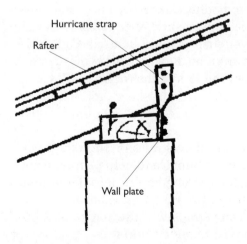

Hurricane strap holding down rafter to wall plate

Rafters and the ridge board

The ridge board is usually a 200 mm deep piece of board which holds the two sides of the roof together at the top. When high winds pass over a roof, especially one that is flat, an upward suction is created and this will break apart the two halves of the roof at the ridge board.

To prevent this damage, a collar tie should be placed

between every second or third pair of rafters, which will stop the force of the wind pulling apart the two sides of the roof. Alternatively, a steel strap over the top of the rafter can be used.

Lath to rafters

The laths should be placed no more than 75 mm apart, and where possible, 25 mm × 100 mm timber should be used. The laths should be held to each of the rafters with either one screw or two nail fixings (6–8 mm long).

If the laths are too widely spaced on an existing roof, more can be added by lifting the zinc sheeting.

Zinc sheets and laths

Zinc sheets protect the roof from wind and rain. They should be properly nailed down, particularly at the edge of the roof, using zinc nails or screws and fillets to hold the zinc sheeting (wire nails should not be used). A recommended nailing pattern is one nail at every other corrugation along the eaves and ridges, and one nail at every third corrugation in the centre of the roof. Where there is unboxed gable overhang, there should be a nail or screw at every corrugation.

It is important to use the correct gauge zinc; the zinc sheets should be of 26 gauge (28 or 30 gauge is too thin).

Eaves and gable ends

Damage to the eaves and gable ends often starts because these areas are exposed. Overhangs should be kept as short as possible (less than 45 mm) and board edges and cover strips should be used. Patio roofs should be separate from the main roof or they may blow off together. Boxed eaves will also help prevent the loss of the roof.

Finance

There is credit available at affordable interest rates through the Credit Union or the National Housing Trust in Jamaica to help people restore their roofs. The cost of upgrading and strengthening a roof is a small fraction of the cost of replacing the entire roof after it has been damaged by a hurricane. Using hurricane straps will prevent some storm damage if the cost of a retrofit roof is too much initially.

The National Housing Trust

The National Housing Trust (NHT) is a unique Jamaican institution formed in 1976 in response to the urgent need for an agency that would generate additional funding for housing finance and provide shelter solutions for those in need. Through a unique worker/employer partnership, 2 per cent of the gross wages of workers and 3 per cent of employers' wage bills are channelled into the NHT. The combination of corporate and personal savings has created a pool of funds enabling the Trust to provide affordable housing to the lower income groups, based solely on local funding.

Maintenance

Many roofs fail in a hurricane because they become weak from the rusting of nails or sheeting, the rotting of roof timbers and fascia boards, or as a result of unrepaired leaks or simply old age. Termites and woodworm also attack untreated or unprotected timbers. The regular maintenance of roofs keeps them strong, leak-free and prevents costly repairs or replacement.

Tips for keeping roofs safe

- Inspect the roof after heavy rain and fix leaks as soon as they occur.
- Replace defective timbers and rusted sheeting.
- Paint fascia and edge details to lengthen their lives.
- Check for termite trails or dust piles from the roof and ceiling timbers. Treat thoroughly and quickly if infestation is found.
- Do not use untreated lumber in buildings, unless it is termite resistant, for example, cedar.
- Galvanized sheeting tends to rust in areas subject to sea spray and so a thick gauge aluminium sheeting should be used in these areas, or the roof should be kept painted to resist the corrosive effects of salt.

Further information

Construction Resource and Development Centre
11 Lady Musgrave Avenue
Kingston 10
Jamaica
Tel: +809 978 4061 2/982 1763

Earthquake-proof housing*

In Colombia, the poor build with bamboo while the better-off build with cement. As a result of the earthquake in January 1999, vast areas of middle-class housing collapsed, but the bamboo houses remained standing. Yet despite bamboo's proven resilience in earthquakes, it still lacks credibility as a building material.

Bamboo grows to such magnificent heights and strengths that bamboo forests can produce enough 10-metre poles to build cities. To overcome the prejudice of bamboo being for 'poor' housing, in Colombia, houses are being built to look like high-quality suburban homes with thin concrete-covered wire mesh walls and tile topping. The difference is that the walls, first floors and roof are not supported by concrete but by strong and flexible bamboo that will ride out even the most violent earthquake. The attraction for the government is that a three-bedroom bamboo house costs half the price of a concrete one.

Harvesting and storage of bamboo

There are about 600 botanical species of bamboo in the world. Bamboo is a giant grass which grows at a rate of 13 cm (nearly six inches) a day. In six months, it will measure more than 10 metres and it reaches maturity within three years. It takes between four and five years for the bamboo stem, which is hollow, to become hard and strong enough to hold up a house.

Harvesting should be done in the dry season because the bamboo culms (trunks) have a lower moisture content, making transportation easier and reducing the chance of attack by fungi and rot. Once cut, the plant quickly grows a replacement stem. Only adult culms should be harvested because young culms provide food for the bamboo plant. It is important not to cut too many culms from the same plant, otherwise irreparable damage will occur and eventually the plant will die.

Cut bamboo should be stored under cover to protect it from rain and it should also, preferably, be kept clear of the ground. The ground must be clean, free of any rubbish and termites. Good ventilation is essential. Fresh bamboo standing vertically will dry in four weeks, whereas lying horizontally it takes twice as long. If drying occurs too quickly, the bamboo may crack.

Bamboo as a building material

Building with bamboo is a simple technology which does not cause damage to the environment. It uses local materials, cut from the surrounding renewable bamboo forest, preserved with natural smoke. Nothing is wasted and nothing is imported, which means that local people can build with bamboo themselves.

Bamboo is an appropriate and aesthetically pleasing building material for an earthquake-prone region. A bamboo frame keeps its shape very well in earthquakes because it is flexible and durable. Weight for weight, bamboo is stronger than steel.

Although a house can be built completely out of bamboo (except for the fireplace and chimney), usually it will be combined with other building materials such as timber, clay or roofing sheets, according to their availability, suitability and cost.

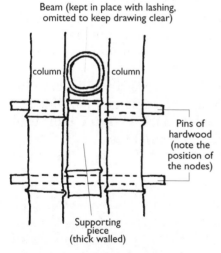

A beam held between two columns of bamboo
© Jules Janssen

A huge structure can be built by joining bamboo together. The bamboo is cut into hollow sections averaging 15 cm in length. To make a joint, a small hole is drilled into a single section which is filled with concrete and then a steel pin is inserted. The joint is attached with another steel pin to the bamboo pole next to it. It is important to place the joint in a bamboo structure either at a node or as near to a node as possible. This technique can be used to hold arches or huge roof spans together. The cement adds strength but does not reduce the flexibility of the bamboo.

Advantages of bamboo

- Bamboo is relatively strong and stiff.
- Bamboo can be cut and split with simple tools.
- The surface of bamboo is hard and clean.
- Bamboo can be grown on a small scale.
- The return on capital is quicker than wood.
- Bamboo structures are flexible in storms and earthquakes.
- Bamboo can be successfully used for the reinforcement of weak soil; for example, to avoid landslides or to strengthen a road.

Disadvantages of bamboo

- Bamboo has low natural durability and needs preservative treatment.
- Fire is a great risk.
- A bamboo culm is not completely straight, it is tapered. The nodes occur at different distances and the prominence of nodes can be a nuisance when the material is being worked.
- Standardization is virtually impossible because of the variation in sizes.

Preserving bamboo

Bamboo is an ideal earthquake-proof material but it needs to be protected against insects. The insects attack the centre of the bamboo because that is the part that holds the most starch and sugar, so it needs to be treated. In the past, bamboo was protected from insects by using chemicals imported from Europe, which were expensive, but now shavings taken from bamboo plants are used to produce a natural smoke insecticide.

Large smoke boxes, which contain bamboo ready-cut for building, are fired with bamboo shavings. These exude a natural pyrolitic acid which protects the bamboo against insect attack. Following this treatment, the bamboo will be insect-resistant for 100 years.

Further information

Technical Adviser
Intermediate Technology Development Group
Schumacher Centre for Technology Development
Bourton Hall
Bourton-on-Dunsmore
Rugby
Warwickshire CV23 9QZ
United Kingdom
Tel: +44 (0) 1926 634400
Fax: +44 (0) 1926 634401
E-mail: infoserv@itdg.org.uk
Website: www.itdg.org

*Information in this section was drawn from Janssen (1995).

Cyclone-resistant health centres

Since the early 1980s, the Indian State of Orissa has been collaborating with the British Government's Department for International Development (DFID) to improve the general health of the population. One area in which this collaboration has been extremely active is the supply and maintenance of new healthcare facilities.

A joint initiative between the local government and DFID has recently completed a series of primary school classrooms using a wide variety of cost-effective construction technologies and innovative designs. Following the success of this initiative, the Government of Orissa entered into a partnership with DFID to develop and build a series of prototype primary health centres, using locally available materials and labour, in three rural areas which required new or improved healthcare facilities. The Government of Orissa supplied the funds and DFID provided the technical assistance.

The project

The technical team given the responsibility for developing the demonstration project was primarily researching hospital designs, alternative building materials and delivery mechanisms in a quest to lower the initial cost of supplying adequate healthcare facilities in the villages of Orissa. They also felt that the situation called for a consideration of all the environmental damage that is created around the supply of more traditional building materials, such as burnt brick and laterite stone blocks.

The project architect, after conducting the area-specific resource mapping exercise (looking at existing healthcare facility designs, local architectural concepts, availability of materials and construction skills, etc.) worked with the Government's Project Management Unit (all medical doctors) and a few selected district medical officers to develop three different primary health centre designs. These were to reflect local aspirations and also rationalize the actual small rural hospital requirements.

Three basic designs – circular, octagonal and a more traditional linear layout – emerged out of this interactive process. Orissa lies along the cyclone-prone Bay of Bengal, so the circular and octagonal forms were chosen for their natural resistance to high winds. A variety of construction technologies and materials were used to construct the primary health centres – under-reamed piles and stub foundation, rat-trap bonded brick and fly-ash block walling, filler slab, brick pyramid, hybrid, fly-ash vault and dome roofing.

The projects display an interesting variety of technologies and built forms:

- the Barikpur project takes inspiration from the traditional Orissan temple form with corbelled brick roofs
- at Panasapada, blocks made from fly-ash generated from a local thermal power station were used both for walling and for the vaulted roofs.

The third building, at Itamati, used fly-ash block masonry and clay tile filler roof slabs.

Other cost-saving building devices, such as arched foundations and openings, have been extensively used in all the buildings. Training and orientation of local engineers and artisans was conducted and the project itself lent great opportunity for 'hands-on' learning. With a significantly higher labour component and minimal dependence on non-local building components, all three projects have contributed substantially to the local economy. Women artisans have played an important role in all the projects.

The durability and structural resilience of the applied technologies were amply demonstrated with all three buildings surviving the super-cyclone of 1999.

Primary health centre at Barikpur

Barikpur is a small village in the Bhadrak District of Orissa. The economy is agriculture based and generally poor. Barikpur is in a black cotton soil belt, which is the major reason for cracks in most of the buildings in the area. The reconnaissance helped in deciding the type of foundation. The local river, Baitarani, offers excellent soil for brick making and, as a result, there are many brick manufacturing units in Barikpur.

Rat-trap bonded brick masonry was chosen for the walls as it was cost-effective and environmentally less damaging than the alternatives at that location. After the first layer of bricks has been laid, a gap is left between the bricks in the remaining courses, reducing the number of bricks and the amount of cement mortar required to build the wall.

The gap in the bricks helps to create thermal insulation, although it is not a good sound insulator. As the mortar is laid on each course, a wooden strip may be used to prevent mortar from falling into the cavities. With this technique, care must be taken to ensure that the wood is not left within the brickwork.

Pyramids were constructed by corbelling or projecting each layer of

BUILDING A SAFE ENVIRONMENT   CHAPTER FIVE   183

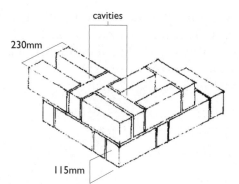

An example of rat-trap bonded brick masonry

bricks from the lower one. A curvilinear profile was introduced by regulating the projections of the corbelled brick courses, ranging from 62.5 mm to 87.5 mm. A parabolic profile is efficient in developing compression, so a corbel of 62.5 mm was used for the first five courses, followed by 75 mm for the next eight courses and 87.5 mm for the last thirteen courses.

The masons could construct two courses per day. The following day, they could stand on these (from the outside) and construct the next two courses, thus eliminating the need for shuttering.

Corbelled pyramids on a square plan were used in the primary health centre, whereas those in the staff quarters were on octagonal plans. The latter is more efficient structurally due to the reduced length of the base, which reduces the secondary bending moment.

The prototype primary health centre and residential quarters for the doctor and the pharmacist were derived from the temples of Orissa. They were designed to demonstrate that upgrading traditional technology and craftsmanship leads to sustainable conservation of a dying art, as well as being cost-effective and socially appropriate.

Fly-ash as a building material

In India, the question of how to dispose of the power industry's fly-ash build-up has been debated for many years, with the construction industry being a major contributor to this debate. Fly-ash, when mixed with lime and gypsum, makes a strong, durable building block, and has become commonly known as FAL-G.

Two of the demonstration primary health centres were built from FAL-G blocks. In Panasapada, the walls were built from FAL-G and the roofs were constructed by using FAL-G blocks to build a continuous circular vault and several individual domes.

Across rural India, it has been difficult to find appropriate materials for shuttering on which to cast a large area of concrete roofing, and this is one of the major reasons for virtually all buildings being delayed. With this in mind, the technical team was keen to develop a roofing system that would minimize the use of shuttering. This resulted in a roof designed and constructed around a series of parabolic arches built with FAL-G blocks to form a continuous vault.

The Panasapada primary health centre was designed and built as a circle around a central courtyard that is open to the sky. The room width between the interior and exterior walls is 3.2 metres.

Construction of the primary health centre at Panasapada

The vault

A truss-type steel mould with a parabolic arch profile was designed for a span of 3200 mm. This had to be light enough to enable easy handling by two masons. Wheels were attached to each end of the mould to allow free movement after decentralizing the individual arches.

The following steps were adopted for the construction of the FAL-G vault:

1. A ring beam was cast along the top of the two external walls.

2. The mould was placed on a 50 mm thick wooden plank, which sat on a bed of wet sand. Levels at both ends of the mould were checked.

3. FAL-G blocks were placed in 1:6 cement and sand mortar on the steel mould. Stone chips were hammered into the mortar joints to ensure contact between the adjacent blocks. A key block was cut to a wedge shape and struck into the crown.

4. The wet sand was removed from under the wooden planks with trowels until the mould came free.

5. The mould was rolled to the next position and the process continued. On average, two masons and three helpers constructed eight arches per day. Because the main building was circular, it was not possible to bond the individual arches together. The primary arches were constructed with staggered FAL-G blocks for interlocking with the adjacent arches. The arches were constructed in a sequence that ensured equal loading around the diameter of the structure.

6. On completion of all the individual arches, the gaps between them were filled with blocks cut to corresponding wedge shapes. As a result, the individually built arches became laterally interconnected and stability of the vault increased considerably.

The main primary health centre roof had 88 individual arches and required 11 working days to be completed.

The domes

The FAL-G domes of the two staff quarters were constructed on an octagonal platform. One continuous band lintel was cast around the walls at a height of 2100 mm from the plinth level. Five courses of corbelled FAL-G masonry were laid on this band to gradually chamfer the corners thus converting the octagons into circles. A reinforced cement concrete ring beam was cast at this level to take the hoop tension exerted by the dome at the base. The domes were constructed in the same manner as the vault but in this case, after the first arch was constructed, the mould was rotated by 90 degrees. The second arch was then constructed exactly as the previous one, and so on.

The primary arches were constructed with staggered FAL-G blocks for interlocking with the adjacent arches. Radial placement of the arches created gaps which were later filled with blocks cut to a wedge shape.

It is thought to be the first time that FAL-G material has been used to construct roofs. In the past, the public has shunned the use of fly-ash-based building products, the main reason being its grey drab colour. The exciting round design coupled with vaulted and domed roofs used in the Panasapada primary health centre has shown that a drab-coloured material does not mean that the end building needs to be dull.

Implementation of these technologies has developed the capacity of the engineers and artisans at village level. It is also worth mentioning that the roof worked out to be approximately 20 per cent cheaper than a reinforced cement concrete slab of similar area. This has encouraged the Health Department of the Government of Orissa to promote the use of FAL-G-based technologies in all future projects.

The huge global demand for more and more physical infrastructure has resulted in an ever-increasing assault upon and consequently further degradation of the environment. The more commonly used construction materials, such as bricks, cement and steel, are produced through systems that are major contributors to this destructive cycle. At the same time, the manufacturing industries are producing millions of tonnes of potentially harmful waste. The complex at Panasapada used approximately 1000 cubic metres of fly-ash, which in its raw form is one of those potentially harmful wastes. This demonstrates that exciting modern structures can be built without constantly making use of materials such as bricks, cement and steel, but by using materials and technologies derived from waste materials that with minimal processing can be both cost-effective and environmentally friendly.

Further information

Sudipto Mukerjee
DFID India
B28 Tara Crescent
Qutab
Institute Area
New Delhi – 010016
India
Tel: +91 11 652 9123
Fax: +91 11 652 9296
E-mail: S-Mukerjee@dfid.gov.uk

For further technical information contact P.K. Das or Peu Bannerjee Das at: pkpeudas@del3.vsnl.net.in

Small-scale brickmaking

The building industry in Zimbabwe is growing fast and new brick houses are being built everywhere. Traditional homes, which are round and built using wattle and daub (pole and dagga), are being replaced by these brick houses. There are standard specifications for bricks in Zimbabwe which effectively prohibit the use of farm bricks (see below) in urban construction. The lowest quality of brick required for general building purposes in towns and cities is called the 'common brick'. Many other countries have similar standards.

Farm bricks

Farm bricks are made using the traditional method of 'slop moulding' brick clay. Slop moulding involves digging up clay and mixing it with water before leaving it overnight to make it ready for moulding. The site should be close to a place where clay can be found easily – for example, large anthills – and near a supply of water. Once moulded, the bricks are laid on the ground to dry in the sun. When dry they are stacked into a clamp, which is a large pile of bricks with firing tunnels built into it (see photograph). The clamp is plastered with mud to provide insulation and burned by lighting wood fires for two or three

Clamp kilns – one being fired, the other dismantled
© ITDG (both photographs)

days. Farm bricks are generally of poor quality, misshapen, underfired, and relatively weak with high water absorption. However, they are adequate for single-storey buildings.

Common bricks

A slop-moulded brick will satisfy the minimum standards required for common bricks, providing the soil used is of exceptional quality and has been carefully prepared. Common bricks must have an average compression strength of 7 Mega-Pascals; they must have a water absorption of less than 15 per cent by weight; and they must resist a specified water spraying test for erodability. Their shape must be regular, their faces smooth and the dimensions uniform. Common bricks are the type most in demand by house builders.

Note: 1 Mega-Pascal = 1 million Newtons per square metre = 145 pounds per square inch.

Crushing

The hammer hoe is a hand tool used for digging clay and can be used for some initial crushing of the soil before it is taken to the manufacturing site. At the site, the soil can be crushed using 'punners', which are heavy metal sections on a vertical shaft used by one person to stamp the soil.

Sieving and weathering

The partly crushed soil is sieved through a screen with a mesh size of 5–6 mm. The soil that does not pass through the mesh is crushed again with the punners and if after this it does not pass through the sieve it is left in weathering heaps for the elements to break down.

Soaking, mixing and tempering

The crushed soil is placed in soaking pits, along with any other materials that make it suitable for brickmaking, such as sand. Water is added and allowed to saturate the soil. The soil must be thoroughly mixed and the cheapest and most effective way of doing this is treading it with the feet, known in Zimbabwe as 'dancing'. The soil is then left to temper in the pits until the water has been completely absorbed and the consistency of the mixture is uniform. Tempering times vary between soils, but generally the longer the better.

Moulding

When the clay has the right moisture content it can be moulded. A good brick can be made in even a simple hand-mould, but the production rate can be increased and the uniformity of the bricks can be guaranteed if a table moulding system is used.

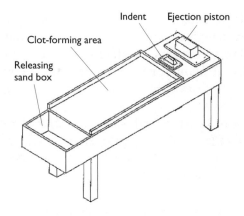
Brick-moulding table

Lumps of clay (clots) are prepared on the table and coated in sand. The mould is placed on the table over a plate with an indent and the clots are placed in the mould. The excess clay is trimmed off and returned to the clot-forming stage, leaving a smooth and consistent finish to the brick. The mould is lifted off the plate and pressed down over an ejection piston, where the brick is left until it is taken to the drying area. The mould is dipped in water, the bottom coated with sand, and the process continues.

Drying

Stacking bricks too high or while they are still wet, or handling them excessively, will result in deformation and damage. The unfired bricks need to be stacked with sufficient airflow between them on a smooth and clean drying surface. They also need to be shaded from the sun to prevent cracking.

Firing

Bricks are fired in clamps after they have been dried. The clamps are flat-topped heaps with steep sides and an insulating layer of mud. The kiln takes approximately a week to build, two weeks to fire and a week to cool, although this is to some extent dependent on the size of the kiln and the weather. Boiler waste is used as fuel for the kiln and is poured into channels between the bricks and smoothed down until it is evenly spread.

Boiler waste as fuel

Coal is produced in Zimbabwe and is readily available in the urban areas. It is used extensively in power stations, as well as in processing plants that use steam or oil for conducting heat – for example, sugar refineries and food processing industries. The efficiency of operation of these plants determines to a large extent how much energy is being extracted from the coal, and how much is being left behind as boiler waste. The ash from inefficient boilers retains a high energy value and burns very cleanly. Boiler ash will always vary in terms of energy value, depending on the efficiency of the power station or processing plant.

Boiler waste is available free from certain industrial and manufacturing plants – particularly Harare's coal-fired power station, which has a problem with disposal and brickmakers need pay only for transport. Consequently, it is a viable energy option for firing clay bricks on a small-scale basis.

The fine particles of the boiler ash are normally sieved out and used as a blending material before moulding, especially in soils with high clay content. This ensures that the brick will burn uniformly, with lower levels of external energy required. These bricks are lighter, as the burnt ash leaves small cavities, and this gives the bricks effective heat insulating properties.

Only the coarse particles are used for firing the clamp. These permit an efficient airflow within the clamp, achieving uniformity and higher temperatures throughout.

The quality of the brick is influenced by the firing temperatures. By using sieved boiler waste as the sole source of energy and the clamp arrangement described above, temperatures of between 950 and 1150°C can be achieved. A rich boiler waste can reach vitrification temperatures, which for most soils are around 1100°C and result in much stronger bricks.

Small-scale brickmakers will build clamps of 20 000 to 30 000 bricks. The larger the clamp, the more energy efficient the firing process. The clamp needs no permanent structure and therefore the only investment costs are hand tools and the land to build the kiln on. The savings derived from using boiler ash are vast compared to commercially sourced coal, because only the transport costs need to be paid and a better quality of brick is produced. The use of boiler waste rather than wood helps reduce the problems of deforestation and subsequent soil degradation.

A clamp kiln showing boiler waste in the channels

Further information

Intermediate Technology Development Group
PO Box 1744
Harare
Zimbabwe
Tel: +263 1 1402896
Fax: +263 4 49041

Making lime cements

Lime is a versatile material produced by burning limestone and is used in two main forms: quicklime and hydrated lime. It is an essential ingredient of many soaps, bleaches and fertilizers, and it can be used in building, road construction, agriculture, water and waste treatment. Essentially, lime is simple and cost-effective to manufacture. It can be produced to an adequate quality in sufficiently small quantities to suit the requirements and conditions in the rural areas of developing countries.

Quicklime

Quicklime is produced by heating any material containing calcium carbonate – usually limestone, but chalk, marble, or sea shells can be used – for several hours, at a temperature of around 1000°C in a kiln. Burning the limestone in this way removes the carbon dioxide in the calcium carbonate and leaves behind calcium oxide and any impurities.

Quicklime must be treated with care because it is a chemically unstable and slightly hazardous product. It can cause serious burns if it comes into contact with moisture and therefore it must be handled carefully.

Hydrated lime

Hydrated lime is produced by adding water to quicklime that has been removed from the kiln. This process is called 'hydration' or 'slaking'. If the quicklime is pure and has been correctly burned, the lumps of lime will chemically combine with the water (so long as it is added in the right proportions) to form a dry powder called calcium hydroxide. This hydrated lime is a more convenient material to handle and use than quicklime.

When more water is added, the hydrate will become a soft and smooth lime putty, which can be stored indefinitely if it is kept wet. If even more water is added the putty will become a milky suspension known as 'milk of lime' which is used in industrial processes where the lime must flow through pipes. When the suspension settles, a clear saturated solution is left above the putty, which is lime water.

Where the proportion of impurities, especially clay, left in the quicklime is very high, the lumps of lime will not readily hydrate and must be finely ground and can then be used only for cement work in building.

Chegutu lime kiln
© ITDG/Kelvin Mason

Lime burning

Lime burners seek to produce the highest-quality quicklime from their limestone at the lowest possible cost. Fuel is one of the major production costs of lime burning and that, coupled with the increasing scarcity of fuel wood and the environmental impact of deforestation, has led the efficiency of the burning process to be assessed according to the amount of fuel it takes to produce a quantity of quicklime. There are various kilns that can be used to perform these tasks and consideration needs to be given to the quality of lime required and the capital, labour and fuel available for the project before selecting the

BUILDING A SAFE ENVIRONMENT   CHAPTER FIVE   193

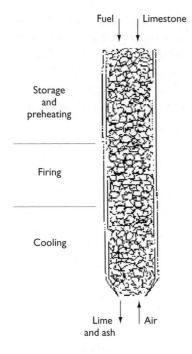

A vertical shaft lime kiln

type of kiln to be used. In Chegutu, the vertical shaft lime kiln was selected.

The vertical shaft lime kiln is designed for small-scale lime producers wanting continuous production of high grade lime with the minimum investment cost. The principle underlying the design of the shaft kilns is the recovery of heat by the upward passage of air through the kiln.

In general, shaft kilns are thermally highly efficient because they use less fuel than other kilns and still produce high-quality quicklime. A chimney is used to provide enough natural draught to draw the air through the kiln and to take the smoke away from the loading area where people work.

A typical small vertical shaft kiln could be a circular brickwork tube, approximately seven metres high, with an internal diameter of around one metre. The height of a shaft kiln should be at least six times the diameter, and these dimensions could be scaled up or down depending on the required output of the kiln and design considerations.

The three zones of a vertical shaft lime kiln

There are no physical boundaries between the zones of a vertical lime kiln; instead they are formed by the careful operation of the kiln. There is a cooling zone at the bottom of the shaft kiln, where the hot lime is cooled by incoming air passing through the quicklime before it is extracted. The firing zone is found about halfway up the kiln and the air that enters it is already warm from cooling down the quicklime. Therefore, less fuel is needed than in traditional box kilns because the hot air is used as a natural resource to heat the limestone to the temperatures required to change it to quicklime. At the top of the kiln, the limestone is stored and warmed up using the waste heat from the firing zone.

The preheated limestone falls gently into the firing zone where the fuel is burning and the limestone begins to change to quicklime. The firing zone must have ample poking holes for monitoring the burning of the lime and to enable it to be readily inspected. A portable ladder on the outside of the kiln should provide access to the poking holes. When the limestone has been completely fired it is broken down to allow the quicklime to fall down into the cooling zone and be cooled before being discharged.

Operation of the kiln

Once the shaft kiln has been lit, limestone that has been broken down into suitably-sized pieces can be fed into the top with a measured amount of fuel. A ramp or a hoist is needed to get the limestone and fuel to the top of the kiln. If a hoist is used, the worker needs to be provided with access to the platform by means of stairs or a ladder.

Efficiency and fuel use

Less fuel is needed in a vertical shaft kiln than in a traditional box kiln, because the hot air is used as a natural resource to heat the limestone to the temperatures required to change it to quicklime.

An important feature of a vertical shaft lime kiln is that, by controlling the airflow through the kiln and by fixing the size and monitoring the composition of the fuel stock, it is possible to obtain a very high quality of lime with few impurities. It is up to the individual lime burner to decide the best size for their limestone pieces within the specified size range and the type of fuel to be used.

As a practical guide, good quality lime can be produced when the limestone to coal ratio is between eight and five to one. This means for every tonne of coal used the kiln will burn about 5 tonnes of limestone and produce between 2.5 and 5.5 tonnes of quicklime. It is possible to use charcoal, wood or other solid fuels, as well as fuel oil, but coal is the fuel used in the vertical shaft kiln at Chegutu in Zimbabwe.

The Chegutu kiln holds about 2 tonnes of limestone and coal when it is full. The daily output of the kiln is between 1.5 and 2 tonnes of quicklime from 3–4 tonnes of limestone. Burning efficiency of this shaft kiln has reached 40 per cent, compared with under 15 per cent for traditional box kilns.

Note: The information provided here is a summary explanation of one of the options available for manufacturers interested in building a lime kiln. The decision to build a vertical shaft lime kiln requires careful consideration of a wide range of situation-specific parameters, many of which are not addressed here.

Further information

Cements and Binders Advisory Service
ITDG
Schumacher Centre for Technology and Development
Bourton Hall
Bourton-on-Dunsmore
Rugby
Warwickshire CV23 9QZ
United Kingdom
Tel: +44 (0) 1926 634400
Fax: +44 (0) 1926 634401

Reducing traffic noise

New locations are needed for housing in the Netherlands and planning policy is now focused on building in and near towns. Although the Netherlands is renowned as Europe's bicycle capital, the volume of motorized traffic on Dutch roads has increased by a third in the past ten years. Increased traffic means increased noise. This is a problem for building companies, which by law have to construct houses to comply with strict noise regulations.

Building at locations near roads and railways forms an essential part of government policy and, over the next few years, the number of locations exposed to high noise levels will increase due to planned infrastructure works. Proposals have been made to reduce the noise problem. Initially, noise barriers, such as embankments and mounds, were built to counter the problem of noise but these require major investment and are often unsightly constructions. Furthermore, the enormous increase in the volume of traffic means that the noise bunds would have to be built ever higher.

Bungawalls

In June 1995, in the town of Utrecht, the fourth largest city in the Netherlands, the first 'bungawalls' were constructed. The project was supported by the local authorities.

The bungawall was created to defeat the noise of the main railway in Holland and was built at a distance of about 30 metres from the railway line. The noise level on top of the wall is about 78 decibels and, in nearly every situation, it is about 50 decibels on the roof terraces of the houses built behind the bungawall.

More recently, in Utrecht, a project consisting of 78 houses has been completed. It is situated close to one of the busiest railway lines in the country and is in the vicinity of a motorway which has approximately 86 000 cars travelling along it each day. Fourteen houses were incorporated into the noise bund and it forms an effective barrier to the noise from trains and traffic.

Along the A2 highway in the village of Neerijinen, 29 bungawalls have been constructed. The housing developed alongside the bungawalls contain rooms that are sensitive to noise, such as the kitchen, attic, bathroom and staircase, and they are designed against the 'blind' and 'deaf' elevation created by the bungawalls.

Construction of the bungawalls

The elevation is situated about 40 metres from the highway but the residents hardly hear the noise of traffic. An earthen wall about 10 metres high and 300 metres long was built to reduce the levels of noise from about 80 decibels to 53 decibels (a reduction of 3 decibels reduces the noise level by half).

The wall is constructed using sand and black soil and is built at a

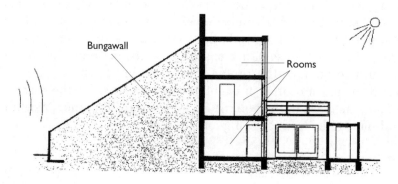

Cross-section of a house incorporated into a bungawall

gradient of 1:2. It is important to use black soil (or clay) because it naturally clings together and therefore will not slide down the slope, which is at an angle of 45 degrees. The concrete wall is covered with a water-resistant layer made from the same materials usually used to construct roofing. Roof insulation 10 cm thick is used to stop the cold coming in. The bungawall is on the west side of the complex, so the rooms, such as the living and sleeping rooms, are built on the east side of the house.

Noise levels

The residents of the houses built along the bungawall have no complaint about the noise levels while they are inside the houses because of the insulation provided by the 10 metre wall and the dike at the back of the house. However, sometimes while the residents are outside on the patio or on the roof terrace during intensive rail traffic the noise can be intrusive.

The noise insulation is slightly more effective if the house is incorporated into the bungawall (see diagram) but this is not absolutely necessary.

Energy demands

The energy demand in these houses is low because the houses are full of natural light which comes in from the roof and the sides facing away from the bungawall. The insulation helps to reduce the noise levels from outside and also provides an effective means of retaining heat in the winter.

Advantages of bungawall housing

Combining a noise bund with housing has many advantages:

- It reduces the need to build housing in ecologically sensitive areas.
- The noise load on existing buildings is reduced.
- Strips of land otherwise unsuitable for housing are used beneficially.
- They have attractive town planning and architectural features.

Local competition for buying the houses incorporated into a noise bund has been immense because people are keen to move into them. The prices of these properties are now at the top end of the housing market.

Further information

Roel Slagter
Wilma Bouw Engineering
PO Box 8591
3503 RN Utrecht
The Netherlands
Tel: +31 30 291 0102
Fax: +31 30 296 7853

Bibliography

Power

Flavin, Christopher (1994) *Power Surge.* Earthscan, London, ISBN 1 85383 205 7
Investment opportunities and environmental problems are pushing the world towards more efficient, decentralized and clearner energy systems. A clear and accessible outline to the massive changes ahead in transport, the home and society.

Goldemberg, Jose (1996) *Energy, Environment and Development.* Earthscan, London, ISBN 1 85383 368 1

Hislop, Drummond (Ed) (1992) *Energy Options: An introduction to small-scale renewable energy technologies.* ITDG Publishing, London, ISBN 1 85339 082 8
Renewable energy can present a baffling array of options to aid agency managers, government officials and advisers. This publication contrasts the relative merits of biomass, solar, hydro and wind power, as well as detailing some direct applications.

Hulscher, Wim and Peter Fraenkel (Introductions) (1994) *The Power Guide: An international catalogue of small-scale energy equipment.* ITDG Publishing, London, ISBN 1 85339 192 1
Dealing with renewable energy sources (wind, sun, water and biomass) this book catalogues small-scale energy equipment (up to 250 kW) and provides information on hundreds of products from almost 500 manufacturers and suppliers in over 40 countries.

Hurst, Christopher and Andrew Barnett (1990) *The Energy Dimension: A practical guide to energy in rural development programmes.* ITDG Publishing, London, ISBN 1 85339 074 7
Helps planners identify energy needs early on so they can be integrated into project design. Summarizes key issues, with detailed discussion of energy options, and the dynamics of supply and demand; plus fact sheets and checklists for quick reference.

ITDG (nd) *Power for Living: Rural Energy in Peru.* ITDG, Rugby

Kristofersen, L.A. and V. Bokalders (1991) *Renewable Energy Technologies.* ITDG Publishing, London, ISBN 1 85339 088 7

Biogas

Fulford, David (1988) *Running a Biogas Programme: A handbook.* ITDG Publishing, London, ISBN 0 94668 849 4
   Describes the designs and uses of biogas plants, with technical appendices, for domestic and community plants. Likely economic and social effects of biogas programmes are described from experience, and advice is given on the problems of management.

Gitonga, Stephen (1997) *Biogas Promotion in Kenya: A review of experiences.* IT Kenya, ISBN 9 96696 066 X

Sasse, Ludwig (1991) *Improved Biogas Units for Developing Countries.* Vieweg, ISBN 3 52802 063 6

van Buren, Ariane (Ed) (1979) *A Chinese Biogas Manual: Popularizing technology in the countryside.* ITDG Publishing, London, ISBN 0 90303 165 5
   Uses diagrams and pictures to show how the basic design of the biogas pit can be adapted for construction in different soils, from sandstone to sheer rock, which should encourage other developing countries to embark on their own biogas programmes.

Biomass

Horne, B. (1996) *Power Plants: An Introduction to Biofuels.* Centre for Alternative Technologies, Wales, ISBN 1 89804 924 2

Solar Power

Berman, Daniel M. and John T. O'Connor (1996) *Who Owns the Sun? People, Politics and the Struggle for a Solar Economy.* Chelsea Green Publishing, Vermont, ISBN 1 89013 208 X

CAT (2000) *Tapping the Sun: A Guide to Solar Water Heating.* Centre for Alternative Technology, Wales, ISBN 1 89804 917 3
   How to adapt your domestic plumbing to use the sun's heat. Over 40 000 solar water heating panels are fitted in the UK, and hundreds of thousands worldwide. This introductory booklet takes the mystique out of this subject, tackling problems such as: Will it work in our cold northern climate? Will the costs outweigh the benefits and money saved? How to fit a panel into a central heating or hot water system. The difference between a flat plate and an evacuated tube. Where to buy the necessary parts and get further information.

CAT (1999) *Solar Water Heating: A D-I-Y Guide.* Centre for Alternative Technology, Wales, ISBN 1 89804 911 4
   How to make two types of solar panel, one using commercially available fins that clip on to copper pipes, and one using a central heating radiator. How to

connect them together, and mount and install them. How to make a control device for a pumped system. Where to go for more information. How to cope with the plumbing, wood preserving and soldering.

Foley, Gerald (1995) *Photovoltaic Applications in Rural Areas of the Developing World.* World Bank, Washington, DC, ISBN 0 82133 461 1
Examines the rural energy context within which PV programmes must fit. The first chapter reviews the present position of PV technology and briefly describes the kits and systems commercially available for use in rural areas of the developing world. The second chapter examines how people manage to meet their energy needs in areas of the developing world that remain untouched by conventional rural electrification programmes. The next chapter looks at conventional rural electrification programmes, their merits and their inevitably limited scope. The fourth chapter looks at the potential niches for PVs, and at how they compare in cost and level of service with their competition. A brief review of PV experience to date and lessons learned is given in the fifth chapter, and the final chapter looks at the role of governments and funding agencies.

Hankins, Mark (revised 1995) *Solar Electric Systems for Africa.* Commonwealth Secretariat, ISBN 0 85092 453 7
Describes how anyone, with help from an electrician, can adapt a small solar electric system to their own needs. Includes details on estimating local resources, choosing appliances and technology, wiring principles, and planning and maintenance.

McNelis, Bernard with Anthony Derrick and Michael Starr (1988) *Solar-Powered Electricity: A survey of photovoltaic power in developing countries.* ITDG Publishing, London, ISBN 0 94668 839 7
A survey of the use of photovoltaics in developing countries, in pumping, refrigerators, lighting, rural electrification and agriculture, from a UNESCO report.

Roberts, Simon (1991) *Solar Electricity: A practical guide to designing and installing small photovoltaic systems.* Prentice Hall, ISBN 0 13825 068 5

Rozis, Jean-François and Alain Guinebault (1996) *Solar Heating in Cold Regions: A Technical Guide to Developing Country Applications.* ITDG Publishing, London, ISBN 1 85339 329 0
A technical guide to the design and production of solar installations in regions where heating is an issue of utmost importance. A climate is defined as cold when the average temperature is lower than 10°C, and as severe if temperatures become extreme or vary significantly through the day or over the year. This affects about half the populations of developing countries. This book is written mainly for technicians, architects and designers, with examples of typical installations and how to size them.

Singh, Mandanjeet (1998) *The Timeless Energy of the Sun.* UNESCO, ISBN 9 23103 453 7
The sun is the one energy source that sustains and links all life. Solar power is the origin of many forms of energy, from the wind, to hydroelectric power, to fossil fuels. *The Timeless Energy of the Sun* traces developments in solar technology and its worldwide cultural impact from the Stone Age to the Space Age.

An informative, accessible text and spectacular colour illustrations highlight the ways in which developments in the field of solar technology have improved the lives of people in all sections of society.

Yaron, Gil, Tani Forbes Irving and Sven Jansson (1994) *Solar Energy for Rural Communities: The Case of Namibia.* ITDG Publishing, London, ISBN 1 85339 242 1

The book investigates issues such as the impact on rural development, economic viability and consumer willingness to pay, social acceptance and technical performance of alternative solar energy options. Namibia is the focus of attention here, but the methodology adopted and detailed review of international experience make this book relevant to all those interested in solar energy for developing countries.

Wind power

Barlow, Roy N., Francis Crick, Peter Fraenkel, Anthony Derrick and Varis Bokalders (1993) *Windpumps: A guide for development workers.* ITDG Publishing, London, ISBN 1 85339 126 3

The wind is a renewable energy resource that can never be exhausted, and which avoids pollution, making it one of the most environmentally sound energy options available. This book takes the reader through every aspect of wind energy in a systematic way.

Gipe, Paul (1999) *Wind Energy Basics: A Guide to Small and Micro Wind Systems.* Chelsea Green Publishers, Vermont, ISBN 1 89013 207 1

Hills, Richard L. (1996) *Power from the Wind: A history of windmill technology.* Cambridge University Press, Cambridge, ISBN 0 52156 686 X

Piggott, Hugh (1997) *Windpower Workshop: Building your own wind turbine.* Centre for Alternative Technology, Wales, ISBN 1 89804 927 0

Windpower Workshop explains very clearly the steps involved in setting up a wind system, including a valuable market survey of wind turbines, suppliers and support organizations.

van Dijk, H.J. and others (1990) *Windpumps for Irrigation.* TOOL

This book provides readers with sufficient information on windpumps to judge their utility under various circumstances. Besides technical aspects, the authors discuss issues such as collection of climatic data and economic feasibility.

Water power

Eisenring, Marcus (1991) *Micro Pelton Turbines.* SKAT, St Gallen, ISBN 3 90800 134 X

Micro Pelton Turbines is a manual on the layout, design, manufacture and installation of very small, locally built Pelton turbine plants. This publication is directed to those who intend to design, dimension, build, install and run small Pelton turbines. It provides all the necessary theoretical background, designs and hints on manufacturing and on procedures of installation.

Fraenkel, Peter (1997) *Water Pumping Devices: A handbook for users and choosers* (2nd edition). ITDG Publishing, London, ISBN 1 85339 346 0

Efficient and effective irrigation of the land can have a dramatic effect on the agricultural output and economic well-being of a community, and smallholdings – defined in this book as up to 25 hectares – can be the source of a significant proportion of a country's food production. At the heart of effective irrigation lies the problem of lifting or pumping water. This handbook surveys the water-lifting technologies that are available and appropriate for smallholdings. It is a detailed and practical review of the options, especially for irrigation but also for other purposes, and the costs and general suitability of the different technologies are examined with the aim of enabling farmers and policy workers to make informed choices.

Fraenkel, P., O. Paish, V. Bokalders, A. Harvey, A. Brown and R. Edwards (1991) *Micro-hydro Power: A Guide For Development Workers.* ITDG Publishing, London, ISBN 1 85339 029 1

Harvey, Adam with Andy Brown, Priyantha Hettiarachi and Allen Inversin (1993) *Micro-hydro Design Manual: A Guide to Small-Scale Water Power Schemes.* ITDG Publishing, London, ISBN 1 85339 103 4

The manual provides worked examples throughout so that financial and engineering calculations can be followed in detail. These examples will allow the reader to design his or her own scheme with confidence. The manual has drawn on the extensive field experience of many practitioners in addition to that of ITDG's micro-hydro team.

Inversin, Allen R. (1986) *Micro-Hydropower Sourcebook: A Practical Guide to Design and Implementation in Developing Countries.* NRECA, ISBN 0 94668 848 6

In some countries, local technical expertise has been developed, often by trial and error. For these people, the sourcebook can provide a reference and guide to help them build on their experiences. In other countries, local expertise is not readily available and as consultants can significantly increase the cost of implementing a scheme, they are not always a viable option. In such cases, the sourcebook can serve as a detailed primer for people with basic technical aptitude, covering many aspects of planning for, designing and implementing micro-hydropower schemes.

Ross, David (1995) *Power from the Waves.* Oxford University Press, ISBN 0 19856 511 9

In this book, Ross explains clearly and comprehensively how wave power could work, how experimental stations do work, and the politics and vested interests that have hindered it, and continue to do so. He writes from first-hand experience and with authority – he attended the events that he describes and knows personally the engineers, scientists, politicians and civil servants involved. The book contains an urgent environmental message combined with a detailed technical account, yet it is written in plain English understandable to scientist and layperson. It shows that wave energy is a resource that creates no environmental problems and which cannot be depleted because 'the waves go on for ever'.

Energy Efficiency

Anderson, Victor (1993) *Energy Efficiency Policies*. Routledge, London, ISBN 0 41508 697 3

Christensen, Karen (1995) *Green Home: How to make your world a better place*. Piatkus Books, London, ISBN 1 85750 200 0

Dichler, A. (1994) *Review of the Technical Merits of a Proposed Photovoltaic House in Oxfordshire*. ETSU

Jackson, Frank (1995) *Save Energy, Save Money*. Centre for Alternative Technologies, Wales, ISBN 1 89804 905 1

Roaf, Sue, Manuel Fuentes and Stephany Thomas (2001) *Ecohouse: A design guide*. Architectural Press, ISBN 0 75064 904 6

Roaf, Sue (nd) *21AD/PV: Photovoltaics*. School of Architecture, Oxford Brookes University, Oxford

Viljoen, A. *'Low Energy Dwellings: the significance of embodied energy and energy in use'*, MSc Architecture advanced energy and environment studies 1994–5, UEL School of Architecture Centre for Environment and Computing in Architecture

World Bank (1998) *Energy Efficiency and Conservation in the Developing World: The World Bank's Role*. World Bank, Washington, DC, ISBN 0 82132 317 2

Stoves

Allen, Hugh (1991) *The Kenya Ceramic Jiko: A manual for stove makers*. ITDG Publishing, London, ISBN 1 85339 083 6
  The Jiko, a charcoal-burning stove consisting of a ceramic liner fitted inside a metal case, burns 25–40 per cent less charcoal than the traditional stoves on which its design was based. This book provides guidance on its production and promotion.

Gitonga, Stephen (1997) *Appropriate Mud Stoves for East Africa*. IT Kenya, ISBN 9 96696 064 3
  This publication has over 150 illustrations and is an easy-to-understand guide on production, promotion and dissemination of mud stoves in East Africa. It provides the basic rules of thumb that organizations and communities need to observe in order to improve on the efficiency of traditional fires. It will be of interest to field staff, decision makers in development agencies, project plan-

ners and implementers, community members and promoters of appropriate energy technologies.

Joseph, S.D., Y.M. Shanahan and W. Stewart (1985) *The Stove Project Manual: Planning and implementation.* ITDG Publishing, London, ISBN 0 94668 875 3
Detailed guidelines for project managers interested or involved in stoves programmes. The book draws on ITDG's experience in helping associations to implement stove programmes.

Micuta, Waclaw (1985) *Modern Stoves for All.* ITDG Publishing, London, 0 94668 835 4
Revised edition of this practical survey of low-cost stoves for use in developing countries. Prepared for skilled technicians who will be able to use the drawings as a basis for building stove models, for testing and adaptation to local conditions.

Stewart, Bill N. et al. (1987) *Improved Wood Waste and Charcoal Burning Stoves: A practitioner's manual.* ITDG Publishing, London, ISBN 0 94668 865 6
A manual for those involved in the day-to-day work of stoves projects. The book describes the chief characteristics, both advantages and disadvantages, of 28 types of stove.

Recycling

Ahmed, Rehan, Arnold van de Klundert and Inge Lardinois (1996) *Rubber Waste: Options for Small-Scale Resource Recovery.* TOOL, ISBN 9 07085 735 9
Documents recovery and recycling activities in cities in economically less developed countries; describes how rubber waste is recovered in informal small-scale enterprises and turned into end products ready for use by other small entrepreneurs and general customers. Attention is paid to the various technologies used in rubber recovery. Financial aspects, marketability of products, environmental problems and government policies are also dealt with.

FOE (1998) *Don't Throw It All Away.* Friends of the Earth, London, ISBN 1 85750 200 0
This new edition of Friends of the Earth's popular recycling guide examines the 'throwaway society' and offers positive solutions to its waste problem. It explains what is thrown away, why so much waste is produced, and the environmental problems this causes. And it offers practical suggestions for reducing the amount of waste you and your family produce.

ILO (1993) *Small-scale Papermaking.* ITDG Publishing, London, ISBN 1 85339 189 1
A technical handbook to assist small-scale producers with alternative production techniques – to help them choose and apply those techniques that are most appropriate to local socio-economic conditions.

Lardinois, Inge (1995) *Plastic Waste: Options for small-scale resource recovery*. TOOL, ISBN 9 07085 734 0
Plastic Waste documents recycling activities in cities in economically less developed countries. The publication describes how plastic waste is reprocessed in informal small-scale enterprises and turned into end products or semi-manufactured products ready for use by formal industries. Attention is paid to the various technologies used in plastic recycling. Financial aspects, marketability of products, environmental problems, occupational health and government policies are also dealt with.

Lindsey, Keith and Hans-Martin Hirt (1996) *Use Water Hyacinth! A Practial Handbook of Uses for the Water Hyacinth from Across the World*. Anamed
A practical handbook of uses for water hyacinth from across the world.

McHarry, Jan (1993) *Reuse, Repair, Recycle: A mine of creative ideas for thrifty living*. Gaia Books, ISBN 1 85675 045 0
An up-to-date sourcebook on how to reduce and recycle, how to create new from old, and how to help to fight the great waste problem of the present age.

Packer, Bevill (1989) *Appropriate Paper-based Technology (APT): A Manual*. ITDG Publishing, London, ISBN 1 85339 268 5
The APT manual is intended for motivated and practical people to study and use to build paper-based articles, as well as to enjoy. This second edition caters for a wider public and new needs, with additional models and extra pages of colour photographs. It aims to set down the special techniques and experiences of APT gained over ten years, with ample examples and illustrations. The special needs of rehabilitation workers have also been catered for, with diagrams and instructions for making prototypes of disability apparatus.

Polprasert, Chongrak (1996) *Organic Waste Recycling: Technology and Management* (2nd edition). Wiley, Oxford, ISBN 0 47196 482 4
This book is a guide to the principles and practice of organic waste recycling. The central core of the book presents a broad range of technologies used in the recycling of organic waste materials to produce valuable products such as fertilizer, biogas, algae, fish and irrigated crops. Each recycling technology is described with respect to objectives, benefits and limitations, environmental requirements, design criteria of the process, use of recycled products and public health aspects. It includes new sections on waste minimization and clean technology, application of constructed wetlands and regulatory aspects of waste disposal and recycling. Case studies of successful recycling programmes are included and exercises for solving both theoretical and practical problems are given.

'Technical Brief', *Appropriate Technology* Vol 24 (4), March 1998

Transport

Barwell, Ian with G.A. Edmonds and others (1985) *Rural Transport in Developing Countries*. ITDG Publishing, London, ISBN 0 94668 896 6
An important and wide-ranging survey of transport policies in developing countries, illustrated by nine case studies.

Clark, J.E. and J. Hellin (1996) *Bio-engineering for Effective Road Maintenance in the Caribbean.* Natural Resources Institute, Chatham, ISBN 0 85954 453 2

Dawson, Jonathon and Ian Barwell (1993) *Roads are Not Enough: New perspectives on rural transport planning in developing countries.* ITDG Publishing, London, ISBN 1 85339 191 3
   Traces the evolution of transport theory and policy and the new needs-led approach, with examples from recent studies. The authors suggest areas of intervention to reduce the transport burden on the rural poor.

Dennis, Ron and Alan Smith (1995) *Low-cost Load-carrying Devices: The design and manufacture of some basic means of transport.* ITDG Publishing, London, ISBN 1 85339 265 0
   The major transport task facing most poor people is to move relatively small loads over short distances, usually off-road. This book describes basic transport technologies such as shoulder poles and back-frames, wheelbarrows and hand-carts, carriers and panniers for bicycles, and load-carrying panniers for animals and animal-drawn sledges.

Dennis, R.A. (1992) *Making Wheels: A technical manual on wheel manufacture.* ITDG Publishing, London, ISBN 1 85339 141 7
   A low-cost technology that will enable workshops to set up their own facilities and manufacture a range of wheels from standard steel sections. For those familiar with metalworking techniques. Presented with technical drawings and sketches.

Doran, Jo (1996) *Rural Transport: Energy and Environment Technology Source Book.* ITDG Publishing, London, ISBN 1 85339 345 2
   This source book seeks to raise awareness and to provide information on how rural transport problems might be identified and addressed, focusing on women, since they carry the main burden. The book highlights rural transport activities and needs, considering household and other agricultural transport needs. It also looks at improving local transport infrastructure and establishing transport services. It uses case study material to illustrate approaches and technologies.

Eisenring, Markus (1998) *Electric Vehicles: With Aspects on Developing Countries.* SKAT, St Gallen, ISBN 3 90800 183 8
   This publication comprises five parts. The first section deals with the history of electric vehicles, the possibilities and limits of electric vehicles, their usage and potential market, legal regulations, conditions for electric vehicles in developing countries and solar-powered vehicles. The second chapter reviews EV technology, power and energy calculations, vehicle conception, batteries, battery charging and drive systems. The focus in the third chapter is on activities and the market situation in the industrialized world and in developing countries. EV energy consumption, environmental impacts and life-cycle economics are studied in the fourth chapter. The final chapter presents future prospects for the development of electric vehicles.

Hathaway, Gordon (1985) *Low-cost Vehicles: Options for moving people and goods.* ITDG Publishing, London, 0 946668 802 8
   A pictorial survey of a wide range of low-cost vehicles in developing countries.

It lists, for the planner and those influencing choice, the range of low-cost options available with their advantages, disadvantages and uses.

Heierli, Urs (1993) *Environmental Limits to Motorisation: Non-motorised transport in developed and developing countries.* SKAT, St Gallen, ISBN 3 90800 141 2
This book examines non-motorized transport – bicycles, rickshaws, electric vehicles – as a decisive component in a new strategy for urban development in both developing and developed countries. It provides useful case studies, tables and many illustrations.

Horne, B. (1996) *Power Plants: An Introduction to Biofuels.* Centre for Alternative Technology, Wales, ISBN 1 89804 924 2

Howell, J.H. (1999) *Roadside Bio-engineering* (two volumes: *Site Handbook* and *Reference Manual*). Department of Roads, Kathmandu. Available free from Geo-Environmental Unit, Department of Roads, Barbar Mahal, Kathmandu, Nepal **or** Infrastructure and Urban Development Division, Department for International Development, 94 Victoria Street, London SW1E 5JL, UK.

Mackenzie, James (1994) *Keys to the Car: Electric and Hydrogen Vehicles for the 21st Century.* World Resources Institute, ISBN 0 91582 593 7

Tickell, Joshua (1999) *From the Fryer to the Fuel Tank: The Complete Guide to Using Vegetable Oil as an Alternative.* Eco-logic Books, ISBN 0 96646 161 4
As pollution envelops the world's cities, temperatures on planet Earth rise, and once rich oil fields run dry, researchers scramble to find solutions to the impending transportation crisis. But the fuel of the future may be hidden in places nobody thought to look. In *From the Fryer to the Fuel Tank*, expert Joshua Tickell unveils the problems with our fossil fuel economy and offers a surprisingly simple solution: cheap, clean-burning vegetable oil. This book provides concise, easy to understand instructions for running a diesel engine on vegetable oil. Packed with photos, graphs and diagrams, it contains all the information you need to become independent of fossil fuels forever.

Building

Aysan, Yasemin, Andrew Clayton, Alistair Cory, Ian Davis and David Sanderson (1995) *Developing Building for Safety Programmes: Guidelines for organizing safe building improvement programmes in disaster-prone areas.* ITDG Publishing, London, ISBN 1 85339 184 0
Summarizes the basic principles to be considered in the planning and implementation of community-based building improvement programmes for small dwellings in disaster-prone areas and includes case studies illustrating suggestions made.

Bee, Becky (1997) *The Cob Builders Handbook: You can hand-sculpt your own home.* Chelsea Green, Vermont, ISBN 0 96590 820 8

A handbook aimed at encouraging the rebirth of natural building. It is written for people with or without building experience and is full of easy-to-ollow diagrams and instructions. Becky Bee has built natural structures in the United States, Australia, New Zealand, Central America and Samoa. Her company, Groundworks, has been at the forefront of the cob revival: building, teaching workshops, and hosting natural building symposiums.

Billett, Michael (1998) *Thatching and Thatched Buildings.* Robert Hale, ISBN 0 70906 225 7
This definitive book on thatching and thatched buildings in rural Britain constitutes an authoritative reference guide. It contains masses of practical information and advice for all those who live in thatched houses or who are contemplating buying one. Michael Billet guides the reader through the development of thatching through the ages, the materials used, and the various types of buildings covered by thatch that can still be seen today. Also included is a chapter that gives advice on costs, including the maintenance and insurance.

Bonner, Roger R.M. and P.K. Das (compilers) *Vidyalayam – Cost Effective Technologies for Primary School Construction.*

Clayton, Andrew and Ian Davis (1994) *Building for Safety Compendium: An annotated bibliography and information directory for safe building.* ITDG Publishing, London, ISBN 1 85339 181 6
Compendium of key publications, organizations, information sources and funding agencies for building improvement programmes. Over 100 selected publications and institutions.

Coburn, Andrew, Richard Hughes, Antonois Pomonis and Robin Spence (1995) *Technical Principles of Building for Safety.* ITDG Publishing, London, ISBN 1 85339 182 4
Details the basic principles to be considered in the planning and implementation of community-based building improvement programmes for small dwellings in disaster-prone areas. Includes sections on earthquakes, flood and wind-resistant construction.

Davis, Malcolm (1994) *How to Make Low-cost Building Blocks: Stabilized soil block technology.* ITDG Publishing, London, ISBN 1 85339 086 0
With the right soil, correctly prepared and compressed, it is possible to halve the amount of cement for blockmaking. In areas where cement is difficult to obtain or expensive, such a saving can be a real boon. This illustrated manual gives, on the strength of successful rural development work, straightforward methods for testing soil and determining quantities of material required. It shows how to manufacture and use strengthened blocks from local soil from initial planning to the finishing touches. There are also details of organizations and literature that can provide further information.

Dudley, Eric and Ane Haaland (1993) *Communicating Building for Safety: Guidelines for methods of communicating technical information to local builders and householders.* ITDG Publishing, London, ISBN 1 85339 183 2

Presents the principles of communicating the information needed for building improvement. Covers the uses of different media to convey information and describes the use of graphic design for education.

Easton, David (1996) *The Rammed Earth House: Rediscovering the most ancient building material.* Chelsea Green, Vermont, ISBN 0 93003 179 2

This book is an eye-opening example of how the most dramatic innovations in home design and construction frequently have their origins in the distant past. By rediscovering the most ancient of all building materials – earth – forward-thinking home builders can now create structures that set new standards for beauty, durability and efficient use of natural resources.

Rammed earth construction is a step forward into a sustainable future, when homes will combine pleasing aesthetics and intense practicality with a powerful sense of place. Rammed earth homes are built entirely on site, using basic elements – earth, water, and a little cement. The solid masonry walls permit design flexibility while providing year-round comfort and minimal use of energy. In this book, David Easton shares the gift of his hard-won secrets for making this age-old technology viable today.

Farrelly, David (1996) *The Book of Bamboo: A Comprehensive Guide to this Remarkable Plant, its Uses, and its History.* Thames & Hudson, London, ISBN 0 50027 911 X

Introducing us to the oldest, most remarkable resource on the planet, this is a practical and historical guide to a versatile wood. Both sustainable and plentiful, bamboo has been used for thousands of years to make a vast array of items – from basic necessities such as clothing and housing, to more exotic and luxurious objects like acupuncture needles, sedan chairs and musical instruments.

Guillaud, Hubert, Thierry Joffroy and Pascal Odul (1995) *Compressed Earth Blocks Volume 2: Manual of Design and Construction.* Companion volume to Rigassi (1995). Vieweg, ISBN 3 52802 080 6

Volume 2 deals with the use of compressed earth blocks (CEBs) and must interest both designers and builders. Its introduction links the use of CEBs with often age-old traditions, and shows what has been achieved more recently, both in the South and the North, using CEBs for various types of building.

Hall, Nicolas (1988) *Thatching: A Handbook.* ITDG Publishing, London, ISBN 1 85339 060 7

A guide to good quality thatching, describing in words and pictures how to achieve the maximum possible roof-life using either cultivated or naturally occurring materials. Reviews the advantages and limitations of thatch as a roofing technique.

Houben, Hugo and Hubert Guillaud (1994) *Earth Construction: A Comprehensive Guide.* ITDG Publishing, London, ISBN 1 85339 193 X

A comprehensive and illustrated handbook that will be essential reading for anyone involved in construction. Earth is extremely versatile and cheap but users must have a proper knowledge of its real potential in order to use it to its best effect.

ILO and UNIDO (1984) *Small-scale Brickmaking.* ILO and UNIDO, ISBN 9 22103 567 0

This technical memorandum, which forms part of a series being prepared jointly by ILO and UNIDO, is the first of three on building materials for low-cost housing. The object of the series is to acquaint small-scale producers with alternative production techniques for specific products and processes, so as to help them to choose and apply those techniques that are most appropriate to local socio-economic conditions. The memorandum provides detailed technical information on alternative brickmaking techniques and covers all processing stages, including quarrying, clay preparation, moulding, drying, firing and testing finished bricks. The techniques described are mostly of interest to small-scale producers in both rural and urban areas. The processes and equipment are described in great detail, with drawings of equipment and tools that may be produced locally, floor plans, information on labour and skill requirements, materials and fuel inputs per unit of output, and so on.

Janssen, Jules J.A. (1995) *Building with Bamboo: A Handbook*. ITDG Publishing, London, ISBN 1 85339 203 0
This revised handbook brings together the practical experience of engineers working in the field and research programmes testing the properties of bamboo. The book shows how bamboo has been used in different designs in developing countries and describes the varying properties and uses of different types of bamboo. The author shows how bamboo can be harvested, seasoned and jointed to form walls, doors and windows, roofs, ceilings, roof trusses and bridges, and how to weave bamboo.

Jayanetti, Lionel and Paul Follett (2000) *Timber Pole Construction: An Introduction* (2nd edition). ITDG Publishing, London, ISBN 1 85339 502 1
This book intends to be descriptive rather than prescriptive, providing ideas that can be developed into solutions depending on specific circumstances, and a number of case studies show some ideas put into action. This new edition will be a useful guide not only to field practitioners in housing and construction projects but will also be of interest to the non-specialist.

Keable, Julian (1996) *Rammed Earth Structures: A Code of Practice*. ITDG Publishing, London, ISBN 1 85339 350 9
Ramming earth has been a method of construction for centuries in various parts of the world. This technique can produce buildings that are strong, durable, safe and desirable, and because earth is an abundant and cheap resource, rammed earth buildings are often very economical. To achieve the best results the right techniques for selection and testing of soils must be used to protect walls from water damage and shrinkage. This book aims to show how high standards can be achieved and the criteria on which rammed earth structures and building techniques can be judged.

King, Bruce (1996) *Buildings of Earth and Straw*. Chelsea Green, Vermont, ISBN 0 96447 187 6
Explores the details and methods for building durable and safe earth and straw houses. While many technical books can be dry, uninteresting and difficult to read, Bruce King has managed to provide technical information in an accessible and entertaining manner. Although parts of this book will admittedly require engineering training to understand, even the uninitiated builder will find a wealth of usable material here.

McHenry, Paul Graham Jr. (1984) *Adobe and Rammed Earth Buildings: Design and Construction.* University of Arizona Press, Tuscon, ISBN 0 81651 124 1

Norton, John (1997) *Building with Earth: A Handbook* (2nd edition). ITDG Publishing, London, ISBN 1 85339 337 1

This handbook provides practical help in choosing whether and how to build with earth, from soil selection through to construction and maintenance. The techniques described in the second edition – revised and updated – of this book have a focus on achieving good quality results with accessible methods, that can go on being used by rich and poor, and for simple buildings as well as the more sophisticated.

Rigassi, Vincent (1995) *Compressed Earth Blocks Volume 1: Manual of Production.* Companion volume to Guillaud (1995). Vieweg, ISBN 3 52802 079 2

Volume 1 deals with the production of compressed earth blocks (CEBs), both technically and economically, and is particularly suited for producers considering making such blocks commercially. The book considers a range of production technologies, from a manual press – 600 blocks per day – to an automated press produced ten times as many.

Spence, R.J.S. and D.J. Cook (1983) *Building Materials in Developing Countries.* Wiley, Oxford, ISBN 0 47110 235 0

Primarily a text for students of architecture and building science in developing countries, this should prove a useful manual for building consultants and engineers with Third World interests. Contents by chapter: 1. Resources, employment and choice of technology 2. Climate, materials and traditional architecture 3. Soil and stabilized soil 4. Brick, stone and masonry 5. Timber and timber products 6. Gypsum, lime and pozzolana 7. Portland and other cements 8. Concrete 9. Ferrocement 10. Composites 11. Roofing materials 12. Economics of building materials choice 13. Social and institutional aspects of choice technology.

Stulz, Roland and Kiran Mukerji (1981) *Appropriate Building Materials: A Catalogue of Potential Solutions.* SKAT and ITDG Publishing, London, ISBN 1 85339 225 1

The book summarizes technical data and practical information from a large number of publications, enabling the reader to identify appropriate solutions for almost any given construction problem in low-cost housing in developing countries, without having to study the volume of literature, which is rapidly increasing every year.

TOOL (1995a) *Rural Building Course Volume 1: Reference.* ITDG Publishing, London, ISBN 1 85339 310 X

Contains information on tools, maintenance of tools, rural building materials, rural building products, tables of figures and a glossary.

TOOL (1995b) *Rural Building Course Volume 2: Basic Knowledge.* ITDG Publishing, London, ISBN 1 85339 315 0

Contains information on basic masonry techniques, basic carpentry techniques and preparation for on-the-job training.

TOOL (1995c) *Rural Building Course Volume 3: Construction.* ITDG Publishing, London, ISBN 1 85339 320 7
   From foundations, walls, doors and windows, to roofs, plaster and render, locks and painting.

TOOL (1995d) *Rural Building Course Volume 4: Drawing Book.* ITDG Publishing, London, ISBN 1 85339 325 8
   A training book for drawing techniques related to the subjects covered in the other volumes of the Rural Building Course, and general drawing training.

Wingate, Michael and others (1985) *Small-scale Lime Burning: A practical introduction*, ITDG Publishing, London, ISBN 0 94668 801 X
   A practical guide to the selection, design and operation of lime-burning plants for small-scale operations. Sections on fuels and raw materials, as well as on the physical and chemical background, and guidance on the methods appropriate to a small scale.

Appendix
The 'Hands On' accompanying videos

Hands On – Food, Water and Finance

Initiative	Originally screened as	Available on video

1 Improved water supplies

Saris as water filters	'Safe Sarees' Saris for Safe Water – Bangladesh	'Sting in the Tale', Hands On 2, Programme 6
Rainwater catchment systems	'Heavens Above' – Kenya	'Waterways', Hands On 2, Programme 1
Hydraulic ram pumps	'Ramming It' - Nepal	'Waterways', Hands On 2, Programme 1
Play pumps	'Play Pumps' – South Africa	'Waterways', Hands On 2, Programme 1
Irrigation for self-reliance	'Pumps, Pipes and Predators'	Earth Report 2, Programme 37
Transporting water	'Water Cigars' – Greece	'Waterways', Hands On 2, Programme 1

2 Better sanitation

Blair ventilated pit latrines	'The Blair Necessity' – Zimbabwe	'Lifting the Lid' Hands On 2, Programme 3
Sanitation support	'Clean Concern' – Sanitation in Jamaica	Earth Report 2, Programme 26
Community sanitation	'The Drain Gang' – Pakistan	'Lifting the Lid', Hands On 2, Programme 3
The way to empty a pit latrine	'Vacutug' – Kenya	'Waste Watchers', Hands On 2, Programme 8

Initiative	Originally screened as	Available on video
Waste management using earthworms	'Worm's Eye' – Ireland	'Lifting the Lid', Hands On 2, Programme 3
Ecological sanitation	'Divide and Spray' – Sweden	'Lifting the Lid', Hands On 2, Programme 3
Large-scale sewage disposal	'Stop the Dump' – UK	Earth Report 2, Programme 23
Wastewater treatment and fish cultivation	'Sewage and Sunshine' – India	'Lifting the Lid', Hands On 2, Programme 3

3 Fruits of the sea and lake

Floodplain fisheries	'A Long Haul' – Indonesia	'Gone Fishing', Hands On 2, Programme 6
Cage aquaculture	'A Cagey Concern' – Bangladesh	'Gone Fishing', Hands On 2, Programme 6
Fertilized fish ponds	'Small Fry, Big Catch', Thailand	Earth Report, Programme 23
Business support for fisheries	'Fishy Business' – Mozambique and 'Fishing for Change – Gill Nets'	'Gone Fishing', Hands On 2, Programme 6 Earth Report 2, Programme 8
Kelp harvesting	'Kelp!' – Ireland	'Gone Fishing', Hands On 2, Programme 6
Seaweed cultivation	'Weed to the Rescue' – Madagascar	'Blood, Sweat and Business', Earth Report 2, Programme 54

4 Small-scale farming

Improving household nutrition	'Less Rice, More Greens' – Vietnam	'Food Works', Hands On 2, Programme 11
Vegetable gardening	'Refugees' – Tanzania for refugees	'Food Works', Hands On 2, Programme 11
Organic farming	'Freedom Gardens' – Malawi	'From the Farm', Hands On 2, Programme 5
Soil stabilization	'Vetiver – a Grassy Solution' – Mexico	Earth Report 2, Programme 38
Agro-forestry	'Shortage to Surplus' – Surviving Mitch – Honduras	'Food Works', Hands On 2, Programme 11
Animal healthcare	'Tough on Grime' – Tanzania	'From the Farm', Hands On 2, Programme 5

Initiative	Originally screened as	Available on video
On-farm income generation	'A Jab in Time' – Vietnam	'Food Works', Hands On 2, Programme 5
Peanut cultivation and processing	'Buttering Up' – Zimbabwe	'Back In Business', Hands On 2, Programme 12
Snack foods as staples	'Snack Attack' – Bangladesh	Earth Report 2, Programme 16

5 Cash crops

Initiative	Originally screened as	Available on video
Fairly traded cocoa	'Pepa de Oro' – Ecuador	'Out of the Forest', Hands On 2, Programme 7
Cashew processing	'Cashing In' – Cashew Processing in Gampaha, Sri Lanka	'What A Difference a Loan Makes', Earth Report 2, Programme 29
Organic olive oil	'100% Virgin' – Olive Oil	Earth Report 2, Programme 33
Vanilla exports	'The Perfect Pod' – Madagascar	'Out of The Forest', Hands On 2, Programme 7
Spice processing	'Spice Girls' Processing in Uruguay	'Do It Herself', Earth Report 2, Programme 12
Organic cotton	'Cottoning On' – India	'From The Farm', Hands On 2, Programme 5
Forest management	'Forest of the Future' – Mexico	Earth Report 2, Programme 28

6 Banking on local enterprise

Initiative	Originally screened as	Available on video
Safeguarding savings	'Safeguarding Deposits' – Madagascar	'Food Works', Hands On 2, Programme 11
Credit union management	'A Mother's Dream' – UK	'What a Difference a Loan Makes', Earth Report 2, Programme 29
Support for village and women's organizations	'On Top of the World' – Pakistan	'What a Difference a Loan Makes', Earth Report 2, Programme 29
Credit for family income generation	'Bamboo Business' – Indonesia	'Out of The Forest', Hands On 2, Programme 7
Youth credit	'Take Five' – Guyana	'What A Difference a Loan Makes', Earth Report 2, Programme 29
Micro-investment	'A Good Return'	'What A Difference a Loan Makes', Earth Report 2, Programme 29

Initiative	Originally screened as	Available on video
New business fund	'Back in Business' – UK	'What A Difference a Loan Makes', Earth Report 2, Programme 29
Loans for phones	'Get Mobile' – Bangladesh	'Back In Business', Hands On 2, Programme 12
Self-help for the homeless	'Another Issue' – Portugal	'Back In Business', Hands On 2, Programme 12

7 Health and safety matters

Safe storage for paraffin	'Safety Caps' in South Africa	Earth Report 2, Programme 11
Improved domestic stoves	'Mirte Stoves in Ethiopia'	'Do It Herself', Earth Report 2, Programme 12
Reducing urban fire risk	'Bomberos 65' – Peru	Earth Report 2, Programme 48
Malaria prevention	'Stop the Bite' – Papua New Guinea	Earth Report 2, Programme 40
Safe blood supplies	'Blood Safe' – Uganda	'Blood, Sweat and Business', Earth Report 2, Programme 54
Medicinal plants	'Cat's Claw' – Peru	'Out of the Forest', Hands On 2, Programme 7
Mercury detection	'A Sniff in Time' – Sweden	'Sting in The Tale', Hands On 2, Programme 4

Hands On – Energy, Infrastructure and Recycling

Initiative	Originally screened as	Available on video

1 Power without destruction

Initiative	Originally screened as	Available on video
Burning biogas	'A Pat Solution' – Nepal *and* 'Where There's Muck' – Germany	'It's A Gas', – Hands On 2, Programme 2 'Who's Got The Power', Earth Report 2, Programme 36
Capturing coal-bed methane	'Managing Methane' – China	'Sting in The Tale', Hands On 2, Programme 4
Micro-hydro power	'Only Connect' – Micro-hydro	'Who's Got the Power', Earth Report 2, Programme 36
Water current turbines	'Alternating Currents' – Peru	'Power to the People', Hands On 2, Programme 9
Harnessing wave energy	'New Wave' – Scotland	'Power to the People', Hands On 2, Programme 9
Wind power	'Wind of Change' – Sri Lanka *and* 'Changing the Current?' – Wind Turbines	'It's A Gas', Hands On 2, Programme 2 'Who's Got the Power', Earth Report 2, Programme 36
Windpumps	'Gone With the Wind' – The Philippines	'Who's Got the Power', Earth Report 2, Programme 36
Solar thermal power	'All Done With Mirrors' – Solar Power – Spain	'Who's Got the Power', Earth Report 2, Programme 36
Wood pulp for power	'Off Piste – Pulp for Power' – Austria	'It's a Gas', Hands On 2, Programme 2

2 Energy-efficient living

Initiative	Originally screened as	Available on video
Reducing household fuel consumption	'A Burning Concern' – Madagascar	'Blood, Sweat and Business', Earth Report 2, Programme 54
Energy-efficient lighting	'Green Lights' – China	'Who's Got the Power?', Earth Report 2, Programme 36
Solar lanterns	'Glowstar' – Kenya	'Power to the People', Hands On 2, Programme 9

220 HANDS ON   ENERGY, INFRASTRUCTURE AND RECYCLING

Initiative	Originally screened as	Available on video
Solar home systems	'Plug and Play' – South Africa	'It's A Gas', Hands On 2, Programme 2
Schools energy saving project	'50:50 The Energy Saving Project' – Germany	Earth Report, Programme 18
The Solar House	'Solar Housing' – UK	'Do It Herself', Earth Report 2, Programme 12
Urban energy policy	'Solar City' – Freiburg Solar Energy – Germany	'Power to the People', Hands On 2, Programme 9
Passive House standards	'The Cepheus Complex' – Austria	'Power to the People', Hands On 2, Programme 9

3 Recycling a valuable resource

Initiative	Originally screened as	Available on video
Household sorting of domestic waste	'Get Sorted' – Recycling in Fredericia – Germany	Earth Report 2, Programme 14
Informal recycling of waste materials	'Waste Busters' – Pakistan	'Waste Watchers', Hands On 2, Programme 8
Community waste collection	'Sweeping Changes' – Bangladesh	'What a Load of Rubbish?', Earth Report 2, Programme 41
Automated recycling of drinks containers	'Cashing In' – Norway	'Waste Watchers', Hands On 2, Programme 8
Cash from cans	'Canning It' – Uruguay	'What a Load Of Rubbish?' Earth Report 2, Programme 8
Accessories from inner tubes	'Inner Style' – UK	'Waste Watchers', Hands On 2, Programme 8
From rags to handmade paper	'From Rags to Riches' – India	'Do It Herself', Earth Report 2, Programme 12
Paper from algae and other wastes	'Pulp Friction' – Italy Algae Paper – Venice	'Waste Watchers', Hands On 2, Programme 8
Rehabilitating water hyacinth	'A Profitable Sentence' – Uganda	'Blood, Sweat and Business', Earth Report 2, Programme 54
Fuel from plastic waste	'Fuel of the Future' – China	'What a Load Of Rubbish?', Earth Report 2, Programme 41
Shipping containers as building materials	'A Clean Conversion' – South Africa	'What a Load Of Rubbish?', Earth Report 2, Programme 41

THE 'HANDS ON' ACCOMPANYING VIDEOS — APPENDIX 221

Initiative	Originally screened as	Available on video
4 Transport for the future		
Intermediate means of transport	'Ease the Jam': Reinventing the Wheel	Earth Report, Programme 20
Bicycle hire	'White Bikes' – The Netherlands	'On the Move', Hands On 2, Programme 13
Electric three-wheelers	'Safa Tempo' Electric Vehicles in Nepal *and* 'Go Electric'	'On the Move', Hands On 2, Programme 13 Earth Report 2
Self-service car rental	'Liselec', France	'On the Move', Hands On 2, Programme 13
Smart cars	'Smart Car', France	'Vehicle to Vogue', Earth Report 2, Programme 13
Hydrogen power	'Hydrogen Cars – Clean Dreams'	'On the Move', Hands On 2 Programme 13
Hybrid power	'Transport of the Future?'	'Vehicle to Vogue', Earth Report 2, Programme 49
Bioengineering to prevent landslides	'Holding up the Himalayas'	Earth Report 2, Programme 7
5 Building a safe environment		
Improved traditional housing	'Changing Lives' – Kenya	'Do It Herself', Earth Report 2, Programme 12
Cost-effective school buildings	'Rat Traps, Domes and Filler Slabs' Schools, Schools, Schools – India	Earth Report 2, Programme 45
Low-cost concrete housing	'Shacking Up' – South Africa	'City Scope', Hands On 2, Programme 10
Earthen architecture	'Mud, Glorious Mud' Earthen Architecture – France	'Vehicle to Vogue', Earth Report 2, Programme 49
Hurricane-resistant roofing	'A Strapping Solution' – Jamaica	Earth Report 2, Programme 31
Earthquake-proof housing	'Bamboozled' – Colombia	'Out of the Forest', Hands On 2, Programme 7
Cyclone-resistant health centres	'Back to the Future' – India	'City Scope', Hands On 2, Programme 10
Small-scale brickmaking	'Bricking It' – Zimbabwe	Earth Report 2, Programme 9

Initiative	Originally screened as	Available on video
Making lime cements	'Cementing Alternatives' – Zimbabwe	Earth Report 2, Programme 9
Reducing traffic noise	'Cut the Noise' – Noisewalls, The Netherlands	Earth Report 2, Programme 27

The videos may be obtained from: TVE, Prince Albert Road, London, NW1 4RZ, UK
Telephone: + 44 20 7 586 5526 Fax: + 44 20 7 586 4866
E-mail: tve-dist@tve.org.uk
TVE online Http://www.tve.org